Historical Background

The Royal Navy's 1749 Articles of War state:

> ***Article XIX. Mutinous assembly.*** *Uttering words of sedition and mutiny. Contempt to superior officers. If any person in or belonging to the fleet shall make or endeavour to make any mutinous assembly upon any pretence whatsoever, every person offending herein, and being convicted thereof by the sentence of the court martial, shall suffer death: and if any person in or belonging to the fleet shall utter any words of sedition or mutiny, he shall suffer death, or such other punishment as a court martial shall deem him to deserve: and if any officer, mariner, or soldier on or belonging to the fleet, shall behave himself with contempt to his superior officer, being in the execution of his office, he shall be punished according to the nature of his offence by the judgement of a court martial.*
>
> ***Article XXXIII. Behaving unbecoming an officer.*** *If any flag officer, captain, or commander, or lieutenant belonging to the fleet, shall be convicted before a court martial of behaving in a scandalous, infamous, cruel, oppressive, or fraudulent manner, unbecoming the character of an officer, he shall be dismissed from His Majesty's service.*

Violations of these Articles set the stage for one of the most famous mutinies in history, great feats of seamanship, and forever changed Polynesian culture and peoples – all over *Artocarpus altilis* or *breadfruit*.

The London Society of West India Planters and Merchants ('The Lobby') was a powerful group of Caribbean plantation owners whose wealth was owed to the thousands of slaves imported to work their sugar and cotton concerns. With an eye to increasing profits, The Lobby had been pressing the Government to introduce breadfruit to the West Indian colonies as an inexpensive alternative to plantain and maize. In 1740 Admiral George Anson noted breadfruit was universally preferred by his company to bread, and its flavour was described by Lieutenant James Cook in 1769 as '… insipid with a slight sweetness resembling that of the crumb of wheaten bread mixed with Jerusalem artichoke'.

The eminent naturalist Sir Joseph Banks came across the plant with Cook in Otaheite (Tahiti) and observed that the breadfruit tree, called *ulu* or *uru* tree, was a staple of life in every sense. It provided food, its bark could be woven into a fine cloth, provided lumber, shelter, and medicine; it passed into Otaheitian legend in which a land where it did not exist was inconceivable. Both Banks and Cook were impressed by its fruitfulness with three crops a year and its ease of propagation by *sucker* (a shoot that grows from the root of the parent plant) which meant the plant did not have to be started from seed. Banks wrote, 'Bread fruit is procured with no more trouble than of climbing a tree and pulling it down. Not that the trees grow here spon-

The breadfruit plant, a print after an original drawn by Sydney Parkinson in 1769. *(© National Maritime Museum, London PAI3935)*

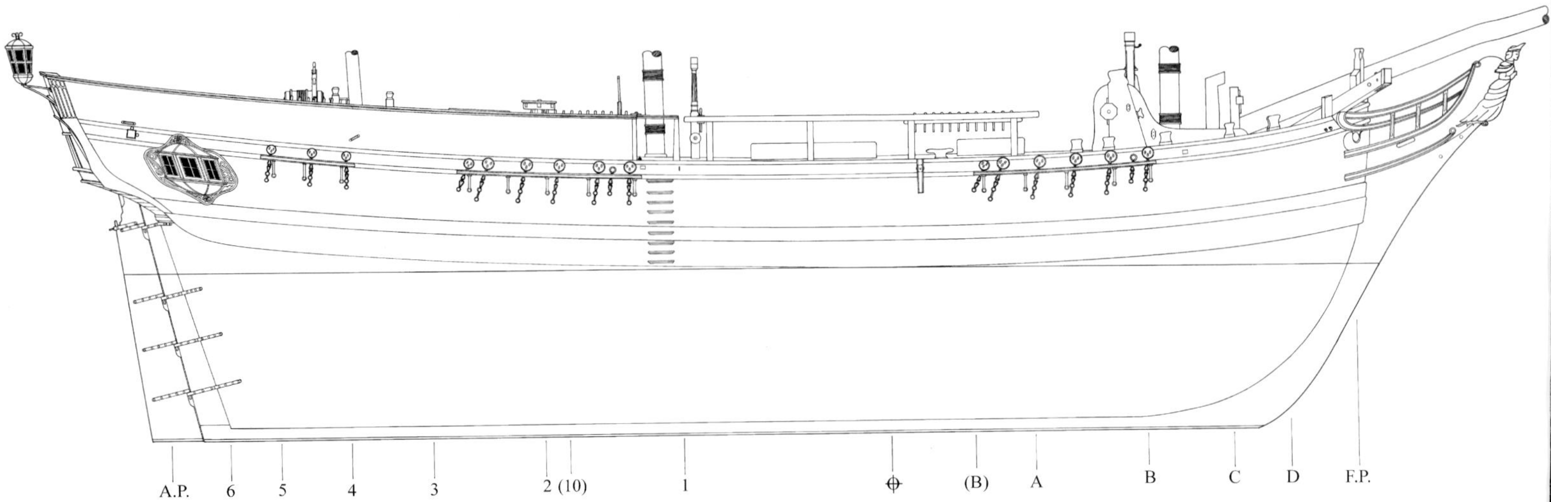

Bethia's Outboard Profile in 1787.

taneously but if a man should in the course of his life time plant 10 such trees which if well done might take the labour of an hour or thereabouts he would as compleatly fulfill his duty to his own as well as future generations.'

The Lobby's interest in breadfruit was also a surreptitious means to undermine a growing abolitionist movement against slavery. The Lobby perverted the ideas of scientific revolution to blunt a growing critique of slavery by showing their benevolence and humanitarianism as slave owners in using botanical science to improve the diet of their slaves.

The Lobby approached Banks because of his studies of the plant and because he would understand the potential it had to feed not just the plantations but a growing British empire across the globe. Banks gained support for an expedition from King George III, and in February 1787, Prime Minister William Pitt the Younger announced an expedition to transport breadfruit was to proceed. On 9 May 1787, Lord Sydney, a Principal Secretary of State, sent a letter to the Admiralty authorising procurement of a suitable ship and the route to Otaheite for the plant's collection.

TABLE 1. *BOUNTY'S* COMPANY

Commander	Lieutenant William BLIGH
Master	John FRYER
Boatswain	William COLE
Gunner	William PECKOVER
Carpenter	William PURCELL
Surgeon	Thomas HUGGAN
Master's Mates	**Fletcher CHRISTIAN**, William ELPHINSTONE
Midshipmen	John HALLET, Thomas HAYWARD, **Peter HAYWOOD, George STEWART** Robert TINKLER, **Edward YOUNG**
Quartermasters	Peter LINKLETTER, John NORTON
Quartermaster's Mate	George SIMPSON
Boatswain's Mate	**James MORRISON**
Gunner's Mate	**John MILLS**
Carpenter's Mate	*Charles NORMAN, Thomas McINTOSH*
Sailmaker	Lawrence LEBOGUE
Master-at-Arms	Charles CHURCHILL
Armourer	*Joseph COLEMAN*
Surgeon's Assistant	Thomas LEDWARD
Captain's Clerk	John SAMUEL
Captain's Servant	John SMITH
Cooper	**Henry HILDEBRANT**
Ship's Cook	Thomas HALL
Butcher	Robert LAMB
Cook's Assistant	**William MUSPRATT**
Able Seamen	**John Adams, Thomas BURKETT,** *Michael BYRN,* **Thomas ELLISON, William McCOY, Isaac MARTIN, John MILLWARD, Matthew QUINTAL, Richard SKINNER, Alexander SMITH, John SUMNER, Matthew THOMPSON,** James VALENTINE, **John WILLIAMS**
Gardener	David NELSON
Assistant Gardener	**William BROWN**

Note: **Bold** = Mutineer; *Italics* = Loyal but detained with mutineers.

THE SHIP

Lord Sydney's letter states:

> 'The Merchants and Planters in His Majesty's West India Possessions have represented that the Introduction of the Bread Fruit Tree into the Islands in those Seas to constitute an Article of Food would be very essential Benefit to the Inhabitants, and have humbly solicited Measures may be taken for procuring some Trees of the Description from the place of their present Growth, to be transplanted in the said Islands.
>
> 'His Majesty is desirous at all Times to promote the Interests of so respectable Body of his Subjects, especially in an Instance which promises general Advantage, ... has thought it fit that Measures should immediately be taken ... for the procuring some Bread Fruit Trees and for conveying them to the said West India Islands, and I am in Consequence to signify to your Lordships His Majesty's Command that you do cause a Vessel of proper Class be stored and victualled for this Service, and to be

fitted with proper Conveniences for the Preservation of as many said Trees as from her Size can be taken on Board, giving command of her to some able and discreet Officer, and when she shall be ready for sea, your Lordships are to direct the said Officer to proceed in her to the Society Islands, situated in the Southern Ocean in the Latitude about 180 degrees South and Longitude about 210 degrees East of Greenwich, where, according to the accounts which are given by the late Captain Cook, Persons who accompanied him during his Voyages, the Bread Fruit Tree is to be found in the most luxuriant state.

'It is proposed that the Vessel which your Lordships are to dispatch upon this Service should proceed around Cape Horn and after she shall have arrived at the Society Islands and as many Trees and plants have been taken on Board as may be thought necessary, to proceed from thence through Endeavour Straits (which separate New Holland from New Guinea) to Princes Island in the Strait of Sunda, or if it should happen to be more convenient, to pass on the North side of that Island, where any Bread Fruit Trees which may have been injured or have died may be replaced by Mangosteens, Duriens, Jacks, Nancas, Lansas, and in short all the fine fruits of that Quarter, as well as the Rice Plant which grows upon dry Land. From Princes Island she should proceed around the cape of Good Hope to the West Indies, calling on her way thither at any Places which may be thought necessary, and deposit one half of her cargo at His Majesty's Botanical Garden at St Vincent for the Benefit of the Windward islands and from thence carry the other half down to Jamaica.

'As from Analogy it appears that the Monsoons prevail in the Seas between New Holland and New Guinea, the easterly Winds will commence in March or April. It is therefore judged proper that the Vessel should sail from hence in the month of September next.

Mr David Nelson and Mr William Brown, Gardeners by Profession, have from their Knowledge of Trees and Plants been hired for the Purpose of selecting such as appear the proper species and Size and it is His Majesty's Pleasure and your Lordships do order those Persons to be received on board the said Vessel and be borne for Wages and Victuals during the Voyage.

'I shall shortly transmit to your Lordships a List of Articles and Trinkets as it is supposed will be wanted to satisfy the Natives of the South Sea Islands for such Trees and Plants as may be obtained from them with an Account of the number Ducats which will be necessary to pay for such Trees and Plants may be purchased from the Inhabitants of Java, in order that the same may be procured and put on Board the Vessel for the purpose of the above mentioned, I shall also forward to your Lordships for your Information a copy of the Instructions which will be prepared for the guidance of the Gardeners, as well in procuring the said Trees and Plants, as in the Management of them after they are put on Board, and your Lordships will give Directions that the Officers and Crew do afford said Gardeners every possible aid and Assistance not only in Collecting of the said Trees and Plants, at the Places before mentioned, but for their Preservation during their Conveyance to the West Indies.'

The same day the Navy Board ordered that a ship not to exceed 250 tons *burthen* (a volumetric measurement of cubic capacity) was to be purchased and outfitted. On 16 May the Navy Board met to consider the vessels on offer:

- *Lynx*, 300 tons, lying at Greenland Dock for £2200 (approximately £367,644 in 2021) by Mr Campbell.
- *Shepherdess*, 270 tons, lying at Pickle Herring Chain, for £2050 (approximately £342,577 in 2021) by Messrs Wellbank, Sharp and Brown.
- *Bethia*, 230 tons, lying at Wapping Old Stairs, for £2600 (approximately £434,488 in 2021), by Messrs Wellbank, Sharp and Brown. (Thought to have been

William Bligh (1757–1817): print by John Conde after a painting by John Russell, 1792. © National Maritime Museum, London PU3227)

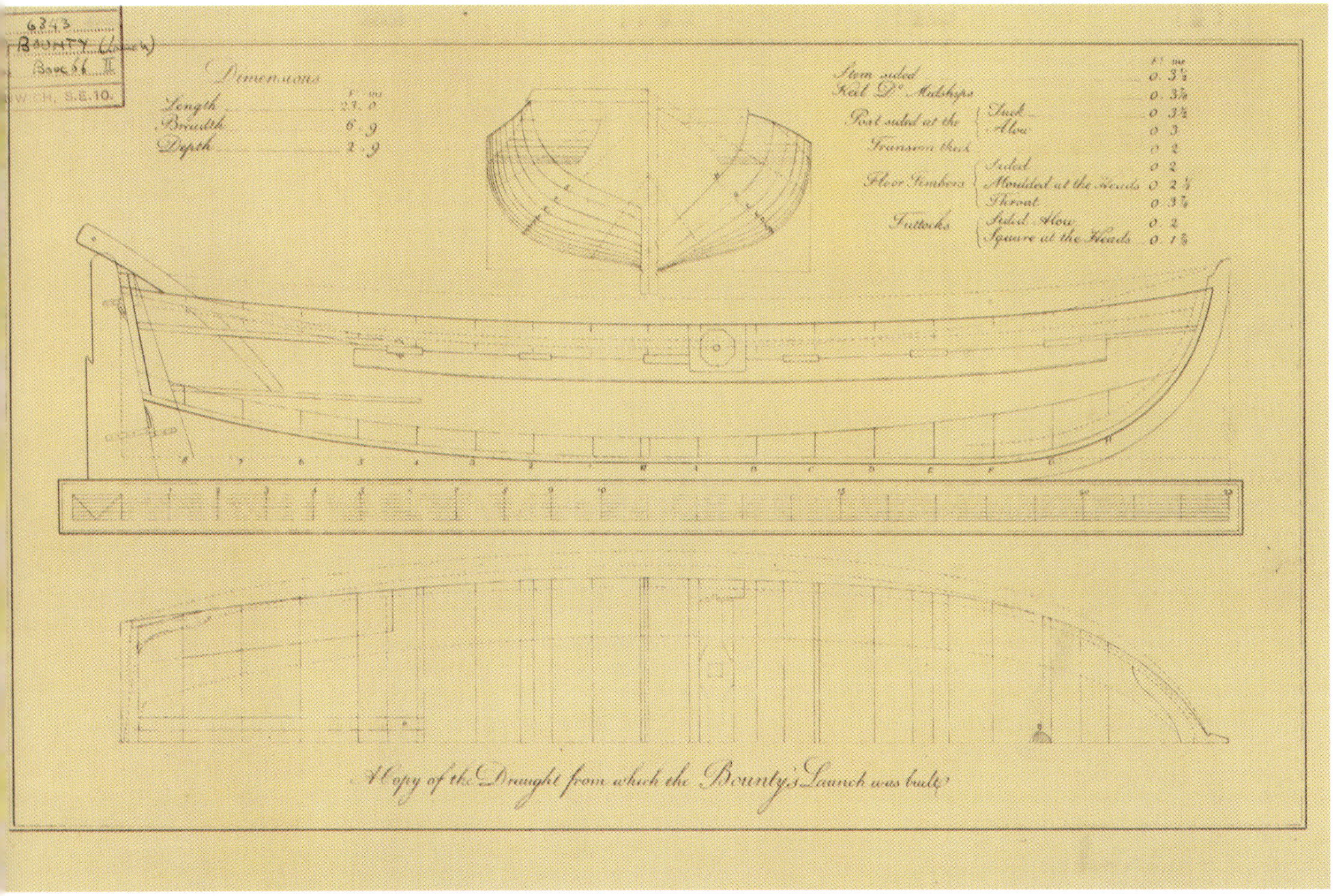

Draught of a 23ft launch as carried by *Bounty*. (© National Maritime Museum, London J2032)

built at Blaydes Yard, Kingston-upon-Hull.)

- *William Pitt*, 240 tons, lying at King Edward Stairs, for £1200 (approximately £200,533 in 2021), by Mr Dawson.
- A new ship, 240 or 250 tons, at Newcastle for £9 10*s* 0*d*. per ton (approximately £1504 in 2021) proposed by Mr Elches.

The *Bethia* was selected and Deptford Yard surveyed her and she was purchased on 9 April for £1950 (£213,829.99 in 2021) with Banks' approval as being suitable for the task. On 31 May the Admiralty directed that she was to be outfitted with four short 4-pounder carriage guns and ten half-pound swivel guns, to sheath her underwater hull in copper with copper fastenings (at the time the hull bottom was sheathed in 1½in thick board), and a complement of twenty men exclusive of officers. The gardener David Nelson was appointed to care for the breadfruit plants and he was to attend Deptford Yard to provide his advice for the outfitting of *Bethia* as a 'floating greenhouse'.

The Navy Board proposed changes to the masts and yards to reduce top weight and to lower the centre of effort of the sails to better handle South Sea weather. The main yard was also to be lengthened, the fore yard reduced in diameter, the topgallant masts reduced in length, and the foremast reduced in diameter. The jibboom was also slimmed in diameter and the cross-jack reduced in both length and diameter from those purchased with *Bethia*.

On 8 June the Admiralty wrote to the Navy Board:

> '…we do hereby desire and direct you to cause the said ship to be registered on the List of the Royal Navy, as an Armed Vessel by the name of Bounty, and established with the number and nature of guns you have proposed, and a complement of 45 men conformable to the scheme on the otherside hereof…'

Three ship's boats were ordered: a 20ft launch, an 18ft cutter, and 16ft jolly boat. Bligh asked if the boats could be exchanged for a 23ft launch and a 20ft six-oared cutter.

On 25 June the Deptford Officers wrote to the Navy Board:

> 'We beg leave to send you the Draughts of the Bounty Arm'd Vessel, as taken off here, The Sheer and Decks are agreeable to the Ship as purchas'd, with the several Contrivances drawn in Red Ink, the Platforms afore and abaft, &c. in Green are propos'd if it meets with your Approval. Sir Joseph Banks propos'd the Bulkhead for the extent of the Garden to be brought to the after part of the after Hatch; to have as many gratings on the Upper deck as conveniently can be to give

Opposite, top: *Bounty*'s sheer and profile draught. (© National Maritime Museum, London J2027)

Opposite, bottom: Alterations to *Bounty's* decks. (© National Maritime Museum, London J2029)

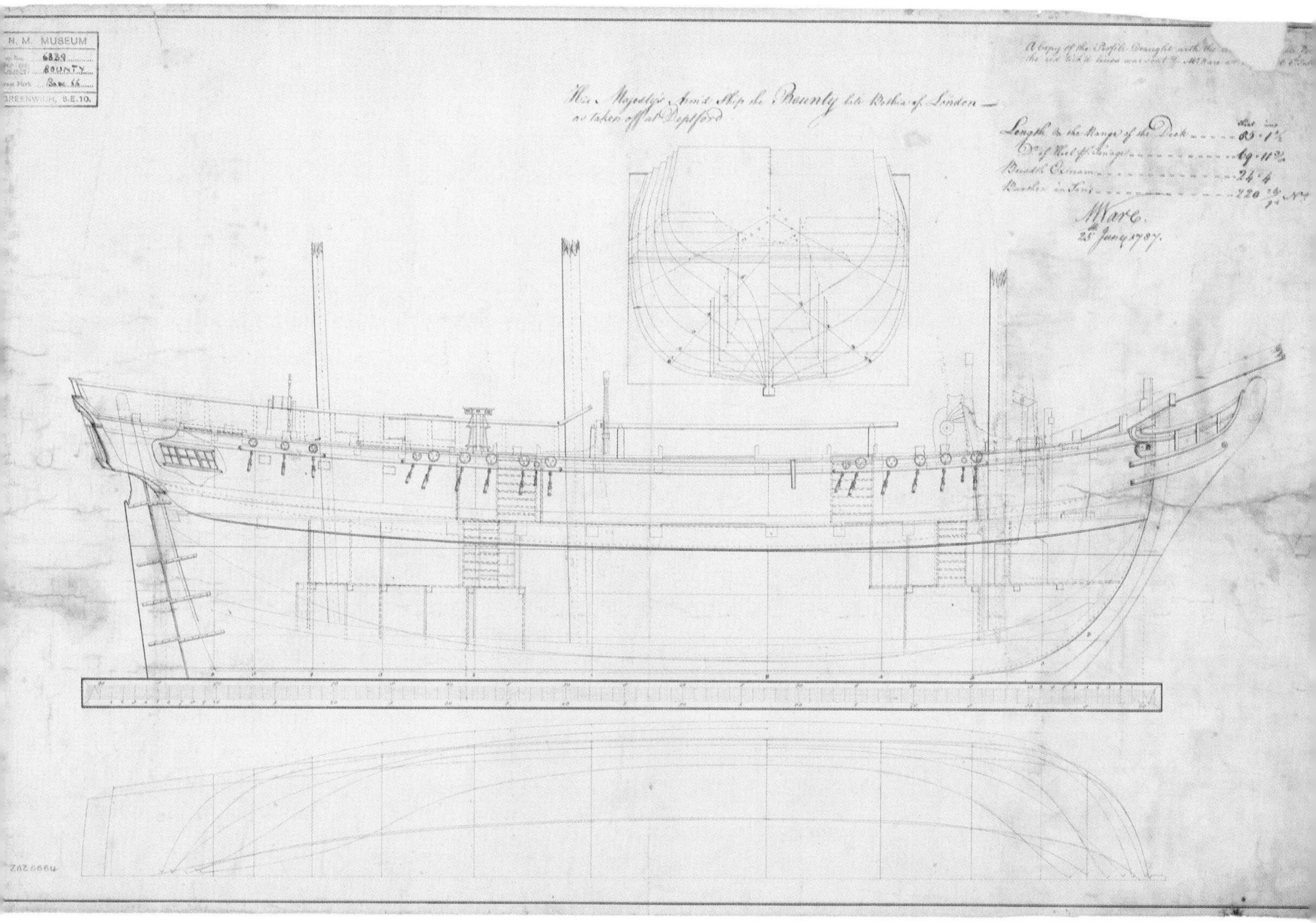
N. M. MUSEUM
6839
BOUNTY
GREENWICH, S.E.10.
His Majesty's Arm'd Ship the Bounty late Bethia of London — as taken off at Deptford
Length on the Range of the Deck .... 85·1½
Breadth Extreme .... 24·4
M Ware
25th June 1787
ZAZ6664

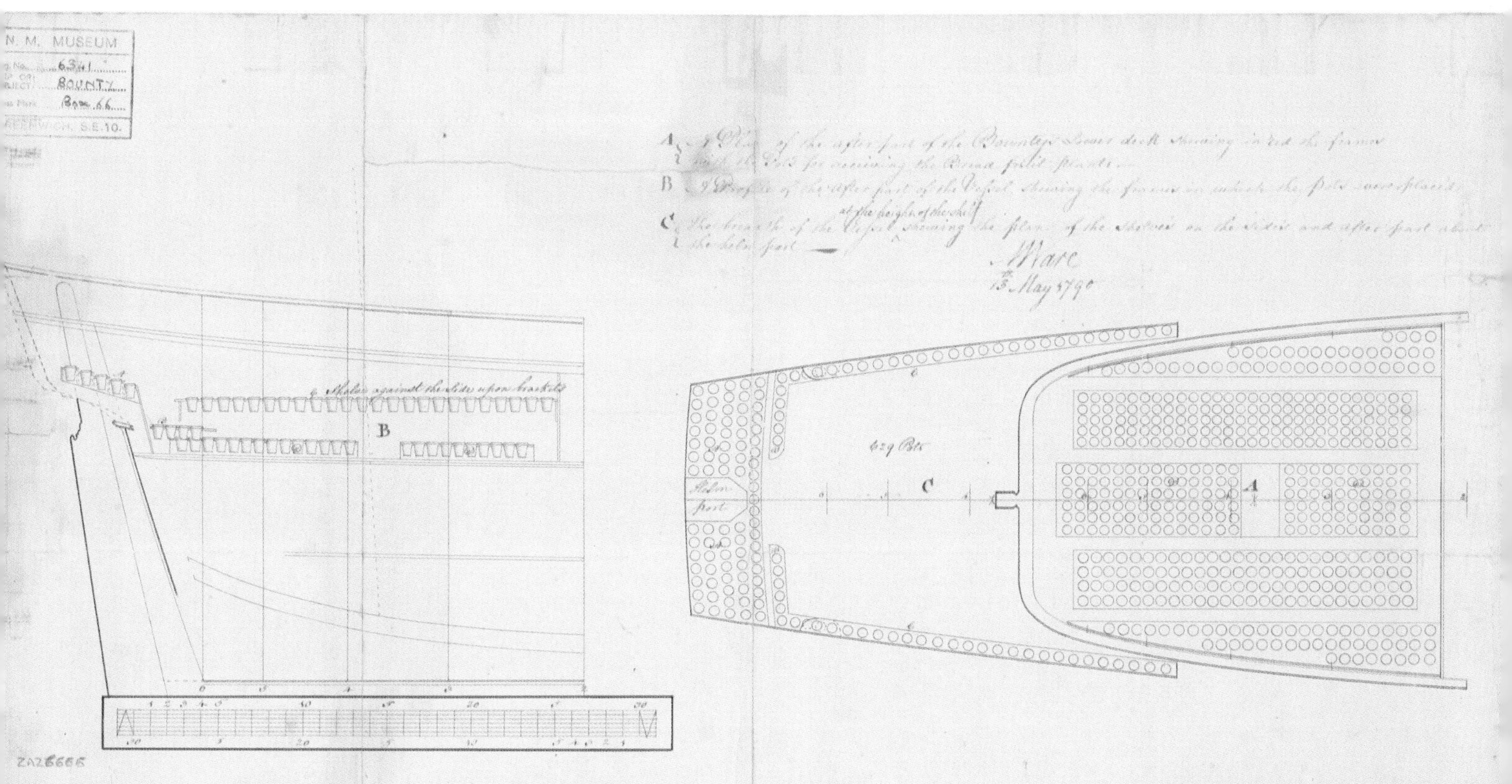
N. M. MUSEUM
6341
BOUNTY
GREENWICH, S.E.10.
M Ware
15th May 1790
B
C
A
ZAZ6666

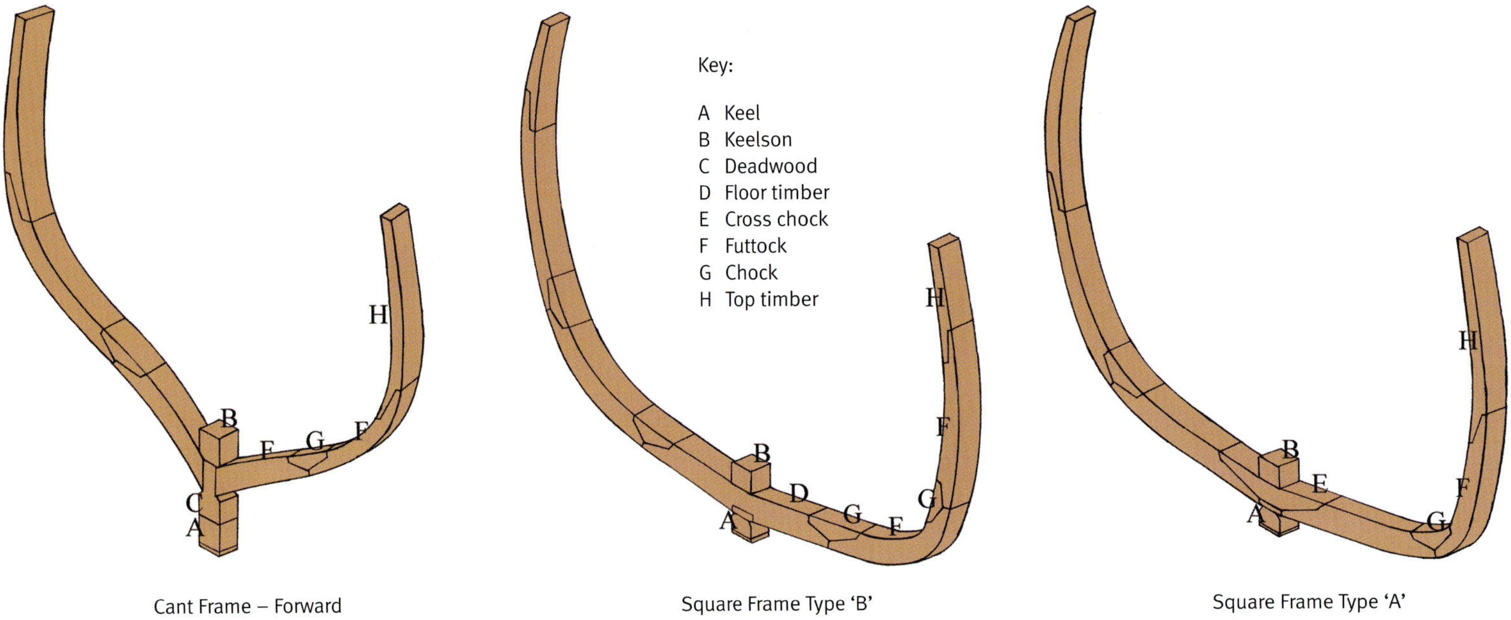

Above: Construction details of *Bounty*'s hull framing.

Opposite: Elevation of *Bounty*'s lower deck garden. Note that the lead lined floor to catch water is shown in grey. The plant pots are shown as terracotta.

Below: Isometric view of *Bounty*'s lower deck garden. The garden is shown whitewashed, but the pot racks could have been unpainted.

air: likewise to have Scuttles through the side for the same reason. He also propos'd the leading the upper part of the Deck, that the water which may drain through the several pots may run to the fore part and be convey'd to a reservoir for the saving it, which being submitted for your Consideration, we pray your Directions thereon.'

In August, Bligh provided his opinion to the Admiralty that the masts should be reduced by the following amounts to better deal with the weather of Cape Horn and the South Seas: mainmast 3ft 10in; main yard 5ft 6in; main topsail yard 4ft 2in; foremast 2ft 4in; fore yard 2ft 5in; fore topsail yard 2ft 6in. He also requested that the boarded solid mast tops were replaced by gratings known as *gratten tops*. The Admiralty agreed. With all of the changes, the cost of fitting out the *Bounty* amounted to £2504 for the hull and £1952 for rigging and stores, giving a total of £4456 (£488,628.93 in 2021). In a memorandum to Banks Bligh wrote, 'The burthen of this ship was nearly 215 tons, her length on deck was 90ft 10in and breadth outside to outside of bends 24ft 3in, a flush deck, and a pretty figurehead of a woman in a riding habit.'

During the voyage Bligh noted the mizzen required shortening by two feet because it was 'too taught and strained the ship' in heavy weather and that he stripped the mizzen of its topmast, topsail, crossjack yards and gaff. However, Bligh did not

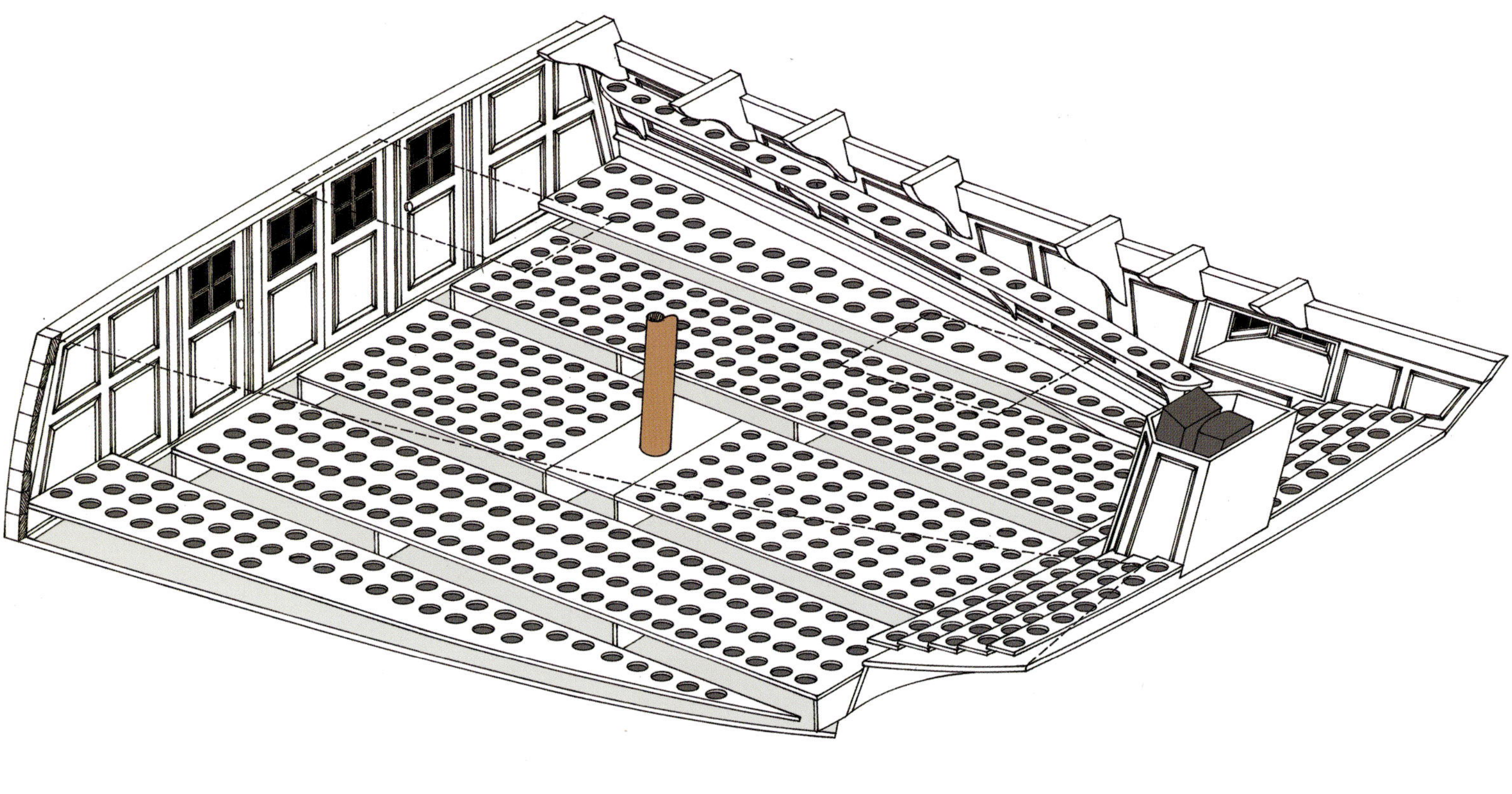

Stern Elevation

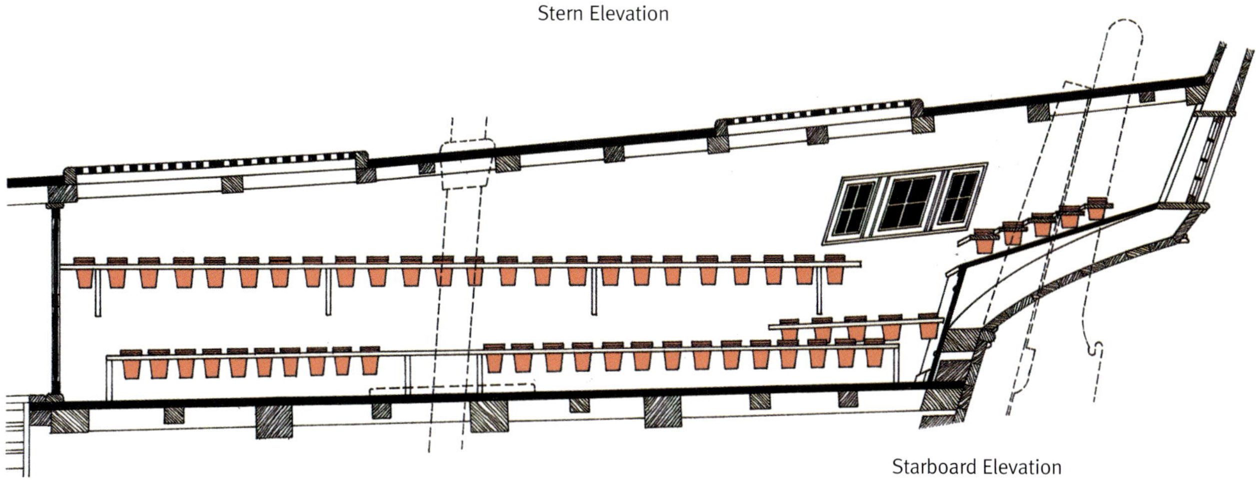

Starboard Elevation

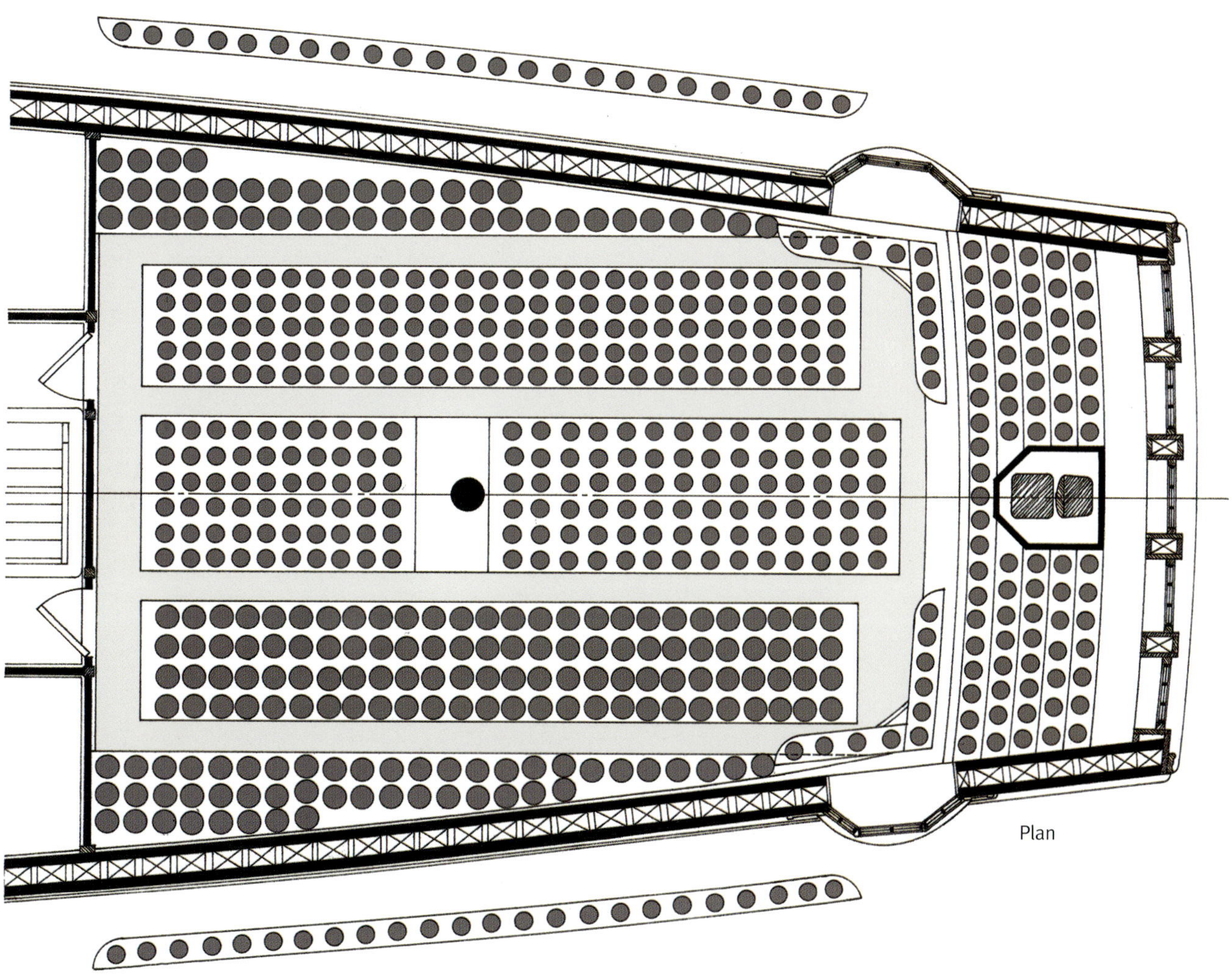

Plan

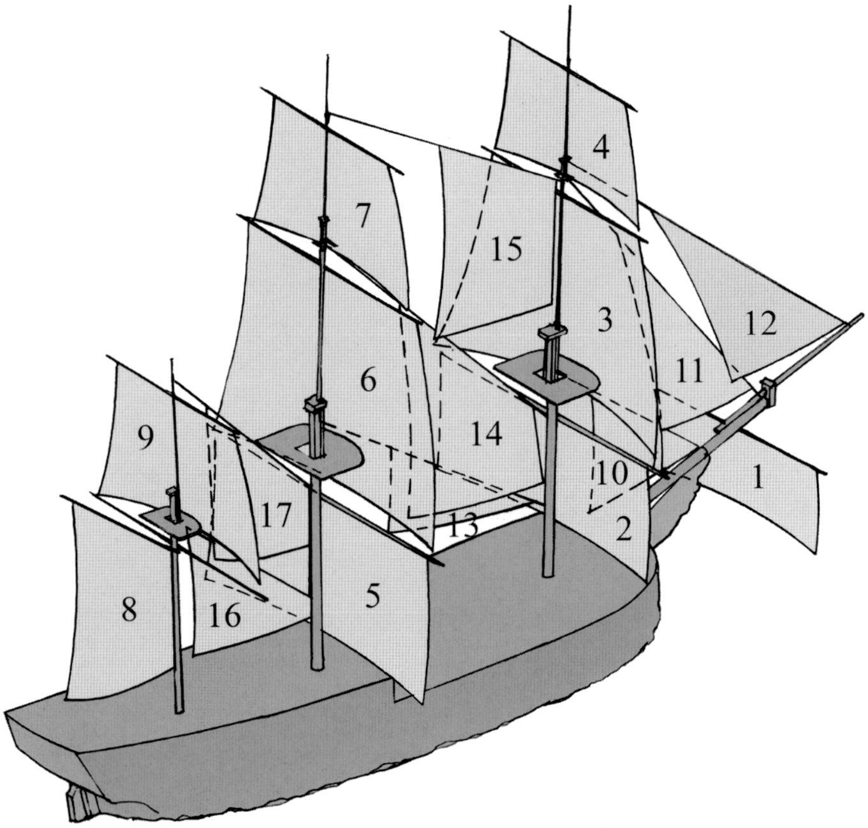

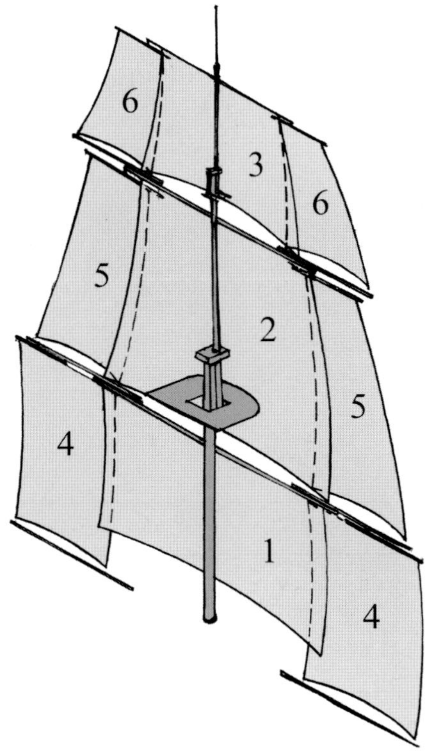

1. Spritsail
2. Fore Course (or Fore Sail)
3. Fore Topsail
4. Fore Topgallant Sail
5. Main Course (or Main Sail)
6. Main Topsail
7. Main Topgallant Sail
8. Mizzen Course (or Mizzen Sail)
9. Mizzen Topsail
10. Fore Stay Sail
11. Fore Topmast Staysail
12. Jib
13. Main Staysail
14. Main Topmast Staysail
15. Main Topgallant Staysail
16. Mizzen Staysail
17. Mizzen Topmast Staysail

Main Mast Studding Sails
(Fore Mast is similar)

1. Main Course
2. Main Topsail
3. Main Topgallant Sail
4. Main Lower Studding Sail
5. Main Topmast Studding Sail
6. Main Topgallant Studding Sail

Bounty's suit of sails.

mention if he took away the boom making it is unclear if it was ever fitted or that it was left in place and unworthy of comment. Bligh's log also noted that the mizzen sail and driver were set like a studdingsail and hoisted to the peak of the gaff. In this case, the sail would be reeved through a block lashed to a pole and run out over the taffrail in lugsail fashion. His notes confirm that the main and fore topsails had four reefs, and that the lower, topmast, and topgallant studdingsails were hoisted on each of the fore and main masts. However, the studding sails were too long and were cut down to make royals of the excess canvas.

The logs note two anchors with wooden stocks and three with iron stocks: two bowers, a sheet, a stream, and a kedge. A launch and a large and small cutter are noted confirming that his request for these boats was granted. The bow of the launch was described as being held in a deck chock that came 'well with the fore hatchway, rather projecting over it'. Bligh described *Bounty*'s performance as an '...excellent sea boat'.

Service History

Bounty's voyage began with the appointment of the 33-year-old William Bligh as Commanding Lieutenant on 16 August 1787. Previously he had served as the sailing master in *Resolution* during Cook's third and final voyage (1776–1780). Owing to *Bounty*'s small size and purpose as a transport, she was rated a schooner and commanded by a Lieutenant who only warranted the courtesy title of 'captain' as commanding officer. The vessel's rating did not entitle her to a section of Marines that traditionally provided protection to the commander and the real authority to enforce his orders.

Bligh was prone to verbal abuse that became worse when he had to enforce his orders through force of personal will. Nevertheless, he was very concerned with the health and welfare of his crew. He changed the navy's traditional watch system of alternating four-hours on and off duty with a three-watch system that allowed for eight hours' rest. He did whatever possible to prevent scurvy with a 'fanatical zeal, continually fuss[ing] and fum[ing] over the cleanliness of his ship and the food served to the crew', and personally dosed any man suspected of the disease with his own remedy of mustard, vinegar, malt, and sauerkraut.

A modern model of *Bounty* in the National Maritime Museum at Greenwich. *(© National Maritime Museum, London I2363_003)*

His antagonist in the mutiny, Fletcher Christian, was paradoxically well-treated by Bligh who received his posting as Master's Mate on *Bounty* on the recommendation of Bligh's wife, Elizabeth Bentham who was well acquainted with Christian's family. Bligh and Christian established a warm relationship and being aware of Christian's desire for a career in the navy promoted him to Acting Lieutenant over his senior John Fryer on 2 March 1788. There is no recorded discord between Bligh and Christian in any log or journal before the mutiny. As for the men, few floggings were recorded and following the example of his mentor Cook, Bligh was sparing of the lash and had expressed hope of a voyage free from such punishment. It is telling that more men remained loyal to Bligh than mutinied. The actual cause of the mutiny remains a mystery and the most popular contemporary accounts were published by relatives (*eg*, Christian's brother, Edward) in a way to preserve the family name and honour by discrediting Bligh.

Modern model of *Bounty* taking on stores. (*© National Maritime Museum, London D4187)*

CAREER

1787
3 September: *Bounty* launched from the dry dock at Deptford.

15 October – 4 November: Sailed to Spithead to take in remaining crew and stores.

29 November: Anchored at St Helens, Isle of Wight. Over the next several weeks contrary winds prevented *Bounty* from reaching the open sea, finally departing English waters on 23 December.

1788
5–10 January: Anchored off Tenerife to take on stores. Bligh writes, 'I have the happiness to tell you my little ship does wonderfully well. I have her now the completest ship I believe that ever swam & she really looks like one fit to encounter difficulties.'

8 February: Crossed equator at 21.50° West and Bligh writes, 'My Men are all good fellows & what has given me much pleasure is that I have not yet been obliged to punish anyone.'

2 March: Bligh makes Christian Acting Lieutenant over his senior, John Fryer.

27 March: Arrival of strong gales and high seas causing *Bounty* to proceed only under staysails. Bligh writes, '... the Storm exceeds anything I had met before and a Sea higher than I had ever seen before, with severe squalls of hail and sleet'.

2 April: *Bounty* approaches Cape Horn. Winds drove the ship back and for ten days Bligh unsuccessfully attempts to round the Horn. The farthest south *Bounty* makes it is 60° 14'.

17 April: Bligh informed his exhausted crew that they would turn and head for the Cape of Good Hope: 'The General Joy in the ship was very great on this Account' and met with three cheers.

24 May: *Bounty* anchored in False Bay. Bligh's letters home emphasised how fit and well he and his crew were and expressed hope that he would receive credit for this.

1 July: *Bounty* departs False Bay for Adventure Bay in Van Diemen's Land.

20 August – 5 September: *Bounty* anchored in Adventure Bay. The ship's company rested, fished, replenished water casks, and felled timber. Encounters with the Aboriginal population were uneventful. The first reports of difficulty between Bligh and his officers emerged when Bligh criticised the carpenter's mate Purcell for cutting billets of wood that were too long. Purcell openly accused Bligh of purposely finding fault and words were exchanged. Purcell was a warrant officer and exempt from flogging, so was sent back to the ship to help with other duties, but even there he openly refused. Bligh being saddled with a small crew and no Marines wrote, 'My directions and presence had little effect' and Purcell's rations were stopped until orders were obeyed. *Bounty* departs on 5 September for New Zealand.

19 September: *Bounty* rounds the southern tip of New Zealand.

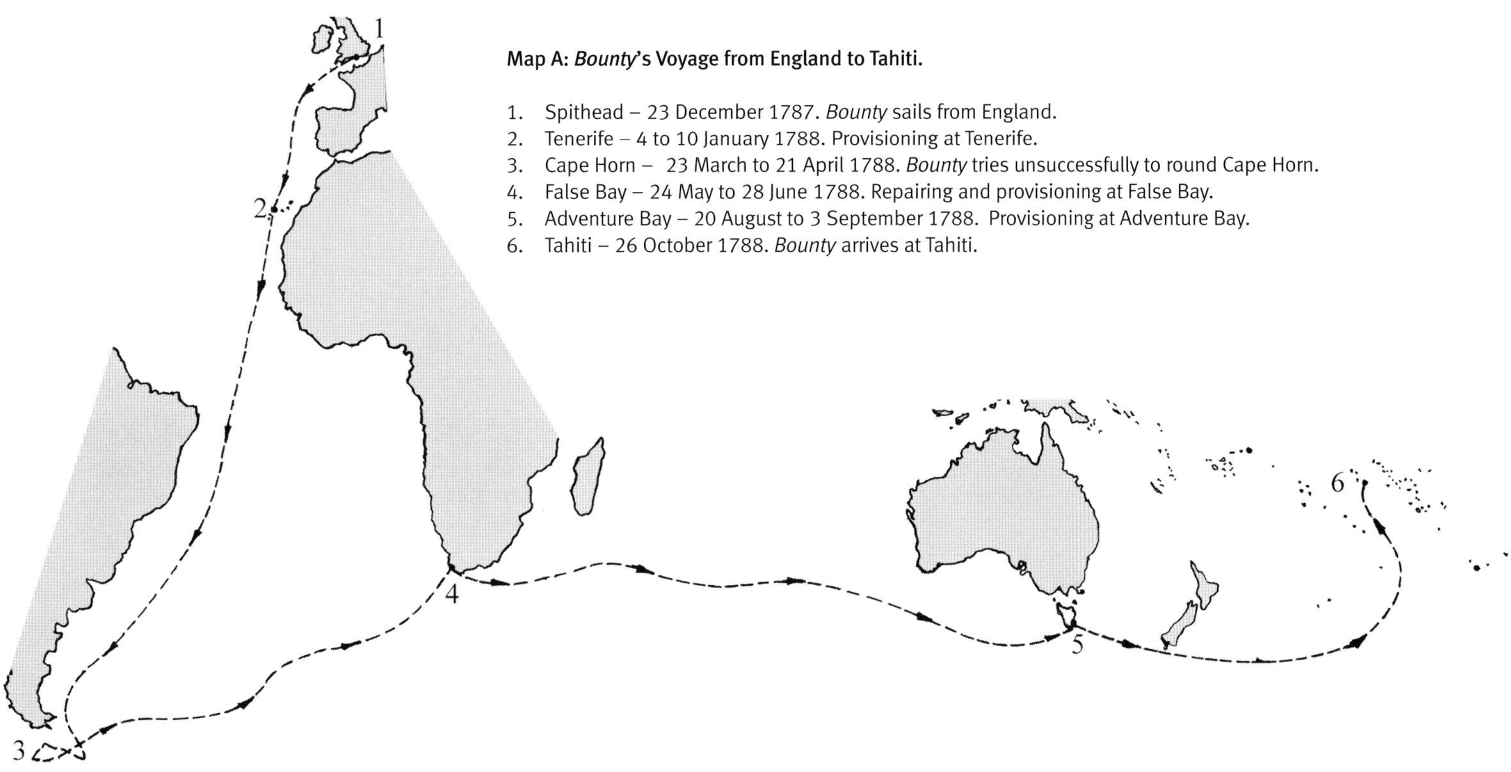

Map A: *Bounty*'s Voyage from England to Tahiti.

1. Spithead – 23 December 1787. *Bounty* sails from England.
2. Tenerife – 4 to 10 January 1788. Provisioning at Tenerife.
3. Cape Horn – 23 March to 21 April 1788. *Bounty* tries unsuccessfully to round Cape Horn.
4. False Bay – 24 May to 28 June 1788. Repairing and provisioning at False Bay.
5. Adventure Bay – 20 August to 3 September 1788. Provisioning at Adventure Bay.
6. Tahiti – 26 October 1788. *Bounty* arrives at Tahiti.

General view of Otaheite from the south-east, oil painting by William Hodges who was on Cook's second voyage. *(© National Maritime Museum, London BHC1935)*

9 October: Fryer demands Bligh provide him with a certificate attesting to his complete competence throughout the voyage so that the Admiralty understood that Christian's promotion over him was favouritism. Bligh summoned the crew and read the Articles of War on which Fryer backed down. Bligh's journal also described the Surgeon Huggan as an '...indolent, unhygienic drunkard.'

19 October: *Bounty* crosses the 25th parallel. The men refused compulsory dance exercise claiming scurvy or rheumatism. Bligh noted no symptoms of scurvy among the men and Huggan who has '... been constantly drunk these four days' continually revised the sick list – adding himself with rheumatism which he changed to 'paralytic affection' (likely referencing the tremors characteristic of alcoholics) to which Bligh responded by confiscating his personal supply of liquor.

24 October: Huggan was ordered to examine every man for venereal disease and reported no cases.

26 October: *Bounty* anchored in Matavai Bay on the northern coast of Otaheite concluding a journey of 27,086 nautical miles (50,163 km; 31,170 statute miles). The bay was well-known to Bligh who had visited there on Cook's second voyage in 1774. They were greeted enthusiastically when they announced they were from 'Pretanee' (Britain) and Bligh immediately sought the co-operation the Paramount Chief King Pomare or Tu-nui-e-a'a-i-te-atua, commonly referred to as Tynah who remembered Bligh and greeted him warmly.

The Otaheitian's enthusiasm for the *Bounty* was attributable to the profitable relations they had enjoyed with Captain Cook. Polynesian society at this time was pre-iron and *Bounty*'s arrival brought iron and other manufactured goods whose possession enabled superior tools and weapons to be created – and the tribe that resourced the most becoming the dominant on the island. Many published accounts reinforce the stereotype that Otaheitian women were sexually promiscuous, but this has been confused with *polyandry* – the practice of having more than one concurrent husband or partner to ensure adequate resources and offspring.

Bligh deliberately did not tell Tynah that Cook had been killed in 1779 at Kealakekua Bay, Hawaii, and forbade any member of the crew revealing that information. This was in part due to the deep esteem the Otaheitians had for Cook and fear that informing them of his death might jeopardise their cooperation. Bligh presented Tynah and his chiefs with gifts and informed them that King George only wished breadfruit plants to which they happily agreed. Clear rules on trade further cemented the relationship between Otaheitans and the British. For example:

> 3rd. Every person is to study and gain the good will and esteem of the natives; to treat them with all kindness; and not to take from them by violent means anything that they may have stolen; and no one is ever to fire, but in the defence of life.

Christian was charged with establishing a nursery that held 600 plants by the end of November. *Bounty*'s extended visit enabled many of the crew to be adopted into an Otaheitian family and protected by family members called *taio*. Medical logs indicated that treatment for venereal infections were received by eighteen officers and men, including Christian. The Otaheitians – according to the British – had long maintained that it was brought by the French. Christian formed a close relationship with a 23-year-old named Mauatua, commonly referred to as Maimiti.

The easy life on Otaheite spawned laxity with increasing numbers of punishments logged. Bligh wrote: 'Such neglectful and worthless petty officers I believe were never in a ship such as are in this.' Sails were allowed to rot, the ship's timepiece was allowed to run down, a boat's rudder and azimuth compass stolen whilst men were on duty. When *Bounty* shifted her moorings to Toahroah Harbour, Fryer allowed the ship to run aground at Oparre.

1789

5 January: Whilst Midshipman Hayward was asleep on his watch, Churchill, Musspratt and Millward deserted by taking a small boat, arms, and ammunition. They were found three weeks later, brought back to the ship and flogged.

27 March: Bligh orders all cats and two dogs disembarked and plants brought aboard.

4 April: Weighed anchor from the harbour at Oparre with a total of 1015 breadfruit plants.

Above: Any portraits of Otaheitians drawn during *Bounty*'s stay did not survive. This is a portrait of Poedua made during Cook's third voyage in 1776 by John Webber. She wears a white drape of tapa cloth made from breadfruit tree bark, her hair decorated with jasmine flowers, and holds a fly whisk in her right hand, a symbol of chiefly rank. *(© National Maritime Museum, London BHC2957)*

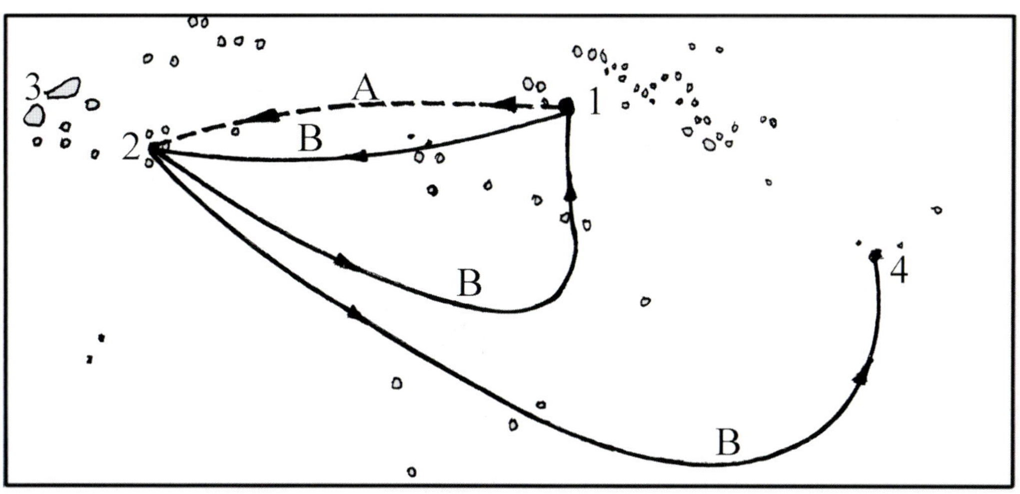

Map B: ***Bounty*** **travels from Tahiti to Pitcairn Island.**

1. Tahiti: 4 April 1789. *Bounty* leaves Tahiti.
2. Mutiny site: 28 April 1789. Fletcher Christian leads a successful mutiny and takes control of *Bounty*.
3. Fiji.
4 Pitcairn Island. After occupying Pitcairn Island the mutineers burn *Bounty* in January 1790. (Precise dates are not known here.)

A. Path of *Bounty* under Bligh.
b. Path of *Bounty* under Christian.

April 5: Bligh and officers took leave of Tynah and his Queen Iddeeah and departed Otaheite. Boatswain's Mate Morrison wrote that everybody 'seemed in high spirits and began already talk of Home'.

22 April: *Bounty* arrives at Nomuka in the Friendly Islands for water and wood before heading into the Endeavour Strait. Bligh had previously visited the island with Cook and was aware the indigenous inhabitants could behave unpredictably. Christian led a watering party equipped with muskets but were under orders their arms should be left in the boat to appear less threatening. On shore, the party was continually harassed causing Christian to give up and returned to the ship – earning Bligh's rebuke that he was 'a damned cowardly rascal'.

27 April: Christian is reported to be in a state of despair, depressed and brooding. His emotional state worsened when Bligh made a comment to Fryer about the size of his store of coconuts. Fryer writes 'Mr. Bligh said they'd been stolen and that it must have been with the knowledge and connivance of the officers. He then questioned each officer as to the number he had bought and, going up to Christian, asked him to state the number in his possession'. Christian answered, 'I do not know Sir, but I hope you dont think me so mean as to be Guilty of Stealing Yours'. Bligh answered, 'Yes, you dam'd Hound I do – You must have stolen them from me or you could give a better account of them – God dam you Scoundrels you are all thieves alike, and combine with the men to rob me – I suppose you'll Steal my Yams next, but I'll sweat you for it you rascals I'll make half of you jump overboard before you get through the Endeavour Straights.' Bligh punished the entire ship's company: 'Stop the Villians Grog, and Give them but half a Pound of Yams tomorrow'.

In despair, Christian considered building a raft with which he could escape to an island. As his discontent became common knowledge among his fellow officers, Midshipmen Stewart urged him not to desert – but Young also assured him that he would have the support of almost all on board if he were to seize the ship and depose Bligh, adding the crew were 'ripe for anything'.

28 April: *Bounty* lay about 30 nautical miles (56 km; 35 miles) south of Tofua. Christian approached Quintal and Martin where he learned the names of others who would join him. With these men Christian gained control of the upper deck and those who questioned his actions were ordered to keep quiet. At about 05:15, Christian went below, dismissed Midshipman John Hallett asleep on the arms chest and distributed arms to his followers. They made for Bligh's cabin and threatened to kill him if he raised the alarm. Bligh called out as loudly as he could, which awoke Fryer who was promptly ordered by the mutineers to 'lay down again, and hold my tongue or I was a dead man'.

Bligh was brought to the quarterdeck by Christian who held the cords binding his hands and brandishing a bayonet. The remaining ship's company were awakened by the noise and joined in the general bedlam on deck. 'Everyone was, more or less, making a noise, either cursing, jeering or just shouting for the reassurance it gave them to do so'. Bligh shouted continually, 'What is the meaning of this violence?' Christian responded 'Hold your tongue, sir. The ship has been taken. Get into the launch, sir.' Bligh responded, 'I have a wife and four children; and you have danced my children upon your knee'. Christian shouted back, 'It's too late! I have been in Hell!' Bligh demanded to be set free, addressing individuals by name, and exhorting the company generally to 'knock Christian down!' Fryer was permitted on deck to speak to Christian, but quickly forced below at bayonet-point. He recounted that Christian told him: 'I have been in hell for weeks past. Captain Bligh has brought this on himself.' It was unclear at this stage who were and who were not active mutineers.

Christian's intent was to cast Bligh, his clerk Samuel and the Midshipmen Hayward and Hallett adrift in *Bounty*'s cutter, but the cutter was not seaworthy so the launch was swung out over the side. The launch had a capacity of ten and over twenty loyalists attempted to board. Christian ordered the loyal Carpenter's Mates Norman and McIntosh and Coleman the Armourer to return to *Bounty* because of their specialised skills. They reluctantly obeyed imploring Bligh to remember that they had remained with the ship against their will. Bligh replied, 'Never fear, lads, I'll do you justice if ever I reach England'.

Samuel saved Bligh's commission papers, the purser's documents, Purcell's tool chest, a compass and quadrant – but not the maps and charts Bligh had made on the voyage. They were also refused the ship's sextant and any timekeepers to help with navigation. The mutineer Morrison gave Bligh his personal sextant and Peckover had a pocket watch to make up the loss. The weight of eighteen men and Bligh and provisions for five days left the launch with only 7 inches of freeboard.

Bligh writes, 'After having been kept some time to make sport for these unfeeling wretches, and having undergone much ridicule, we were at last cast adrift in the open ocean'. At 10:00 they were cut loose and four cutlasses were thrown down into the boat; followed by the breadfruit plants into the sea. As the launch moved away from *Bounty*, Bligh ordered a sail raised to make for Tofua. Christian recognised a

'Just before Sun Rise the People Mutinied, seized me while asleep in my Cabbin tied my Hands behind my back – carried me on Deck in my Shirt – Put 18 of the Crew into the Launch and me after them and set us adrift.' – Lieutenant William Bligh RN, 28 April 1789. Print by Robert Dodd published 2 October 1790. *(© National Maritime Museum, London PY9205)*

search for them was inevitable and decided to head to Tubuai, 450 nautical miles south of Ohtaheite.

28 May: At Tubai, *Bounty* received a hostile reception of war canoes that were dispersed with the 4-pounder guns. The mutineers landed and determined, notwithstanding the Tubians, that the island would be suitable for a defended permanent settlement – but they would require labourers and women.

6 June: *Bounty* returned to Otaheite and told the local Chiefs that he, Bligh, and Captain Cook were founding a new settlement on Aitutaki. The ruse ensured generous gifts of livestock and other goods, and volunteer Otaheitians to build the new settlement.

16 June: *Bounty* returned to Tubuai and began construction of a large moated enclosure christened 'Fort George' after the King. The Tubian chiefs resented their presence and several clashes ensued, leading to discontent among the mutineers. Christian called a meeting on future plans and eight remained loyal to Christian, but sixteen wished to return to Otaheite despite the risk of being found.

22 September: *Bounty* returned to Otaheite to a cool reception. The Chiefs learned from passing ships that the new settlement was a fabrication and that Cook was dead. Christian landed 15 of the 16 men who wanted to return, but for a second time would not allow Coleman to leave. Christian and his loyal party obtained stores but was still short of women. On the pretext of a soirée on *Bounty*, several Otaheitians were invited aboard. The cable was cut kidnapping 26, and as the ship sailed away Coleman went overboard back to shore. Six of the captives were elderly women that were landed on Moorea, 11 miles NW of Otaheite.

The sixteen left in Otaheite formed into three groups. The first group was led by the James Morrison (although Morrison was considered a mutineer, he did not because he maintained that he played no *active* part in the mutiny) and the loyal Tom McIntosh designed and built a schooner christened *Resolution* (after Cook's ship) from breadfruit tree wood. Their plan was to sail to the Dutch East Indies and surrender themselves to confirm their innocence.

Churchill and Thompson who began distilling alcohol from native *ti* plant (*Cordyline terminalis*) and led a lifestyle of drink and debauchery with their Otaheitian followers. The depravity was so extreme that Churchill was murdered by Thompson, who was in turn killed by Churchill's tiao.

The remaining men settled into a quiet life and became fully immersed in Otaheitian culture and lifestyle. They married, had children, had much of their bodies tattooed in the local manner, and learned to speak Otaheitian.

North-westerly view of Pitcairn Island from the south-east painted in 1849. *(© National Maritime Museum, London PAI4613)*

PITCAIRN ISLAND

Christian sailed eastward in search of Pitcairn Island which had been recorded at 25°02'S 133°21'W but found nothing so he zigzagged along the line of latitude.

1790

January 15: Pitcairn was sighted 188 nautical miles (348 km; 216 miles) east of its recorded position. They found it to be a perfect hiding place: inaccessible, uninhabited, fertile and warm. While *Bounty* was anchored in what is now known as Bounty Bay, livestock and goods were shipped ashore.

January 23: *Bounty* was stripped and her hulk burned. Teehuteatuaonoa (Jenny) said that Christian wanted to preserve the ship, but Quintal said 'No, we shall be discovered', so they burnt her.

The mutineers and Otaheitians existed peaceably until mutineers began to consider the Otaheitians as property. The men were treated like slaves and the women were 'passed around from one 'husband' to the other' – when the consorts of John William and John Adams died; Toofaiti and Tinafanaea, who were wives of the Otaheitian men Titahiti, Tararo and Oha, were taken and given to the two mutineers. Their husbands reacted by conspiring to kill the mutineers, but the plot was discovered and Tararo and Oha were murdered sometime during December 1790.

1793 AND BEYOND

By September 1793 the Otaheitians were tired of being treated so poorly and murdered Williams, Mills, Martin, Brown, and Christian. Christian was thought to be killed by Teehuteatuaonoa. Adams was shot, but survived. The Otaheitian men now began fighting amongst themselves over women. Teimua was shot dead by Manarii, then Manarii was shot by McCoy and Quintal. One of the women, possibly Teraura, killed Titahiti and Young shot Niau. At this point, only Young, Quintal, Adams and McCoy remained, along with ten women and their children. Young and Adams assumed a tense leadership and for the next five years they settled into an existence characterised by the poor treatment of the women – who made a failed attempt to leave the island in a makeshift boat that that they could not launch successfully over the rough seas surrounding the island.

Trapped on Pitcairn, the inhabitants were reconciled to their existence and each other. In 1798 this détente was disrupted by McCoy whose distilled alcohol soon brought thoughts of suicide to which he succumbed by drowning himself after binding his own hands and feet. In 1799, an inebriated Quintal threatened fresh murder, forcing Adams and Young to kill him with an axe. Edward Young died of asthma in December 1800. Adams took responsibility for the education and wellbeing of the nine remaining women and nineteen children. He taught basic literacy and Christianity, and kept the island peaceful. Adams was the sole male survivor of the original settlers.

February 1808: The American sealer *Topaz* fell upon Pitcairn. News of the discovery did not reach Britain until 1810, but the Admiralty was preoccupied by war with Imperial France.

1814: The 38-gun Fifth Rates *Briton* and *Tagus* chanced upon Pitcairn and were greeted by a community of 46 people. The Admiralty was informed and decided to take no action.

1829: Adams dies. The community founded by the mutineers continues to thrive today.

1789: THE VOYAGE OF *BOUNTY'S* LAUNCH

29 April: Bligh lands on Tofua but the interactions with the Tofuans soon soured, reputedly from a Tofuan who had a long memory of a personal slight made by Bligh on a previous visit with Cook.

2 May: Bligh sensed an attack was imminent so moved the men and supplies into the launch. The stern line was grabbed by a Tofuan warrior and when John Norton attempted to free it he was immediately stoned to death. Away in the launch all agreed to sail west directly to the Dutch settlement of Cupang, approximately 3500 nautical miles away. Rations would be limited to an ounce of bread and a quarter-pint of water per man. On the 17th Bligh wrote 'our situation was miserable; always wet, and suffering extreme cold ... without the least shelter from the weather'.

28 May: The Great Barrier Reef was sighted and through a narrow gap they landed on a small island Bligh christened 'Restoration Island' due to the abundance of water, oysters and berries.

3 June: The launch heads into the open ocean towards Australia through the Prince of Wales Channel in rough weather.

13 June: The launch arrives at Cupang.

John Adams as drawn by Edward Beechy in 1825. *(© National Maritime Museum, London PAD3414)*

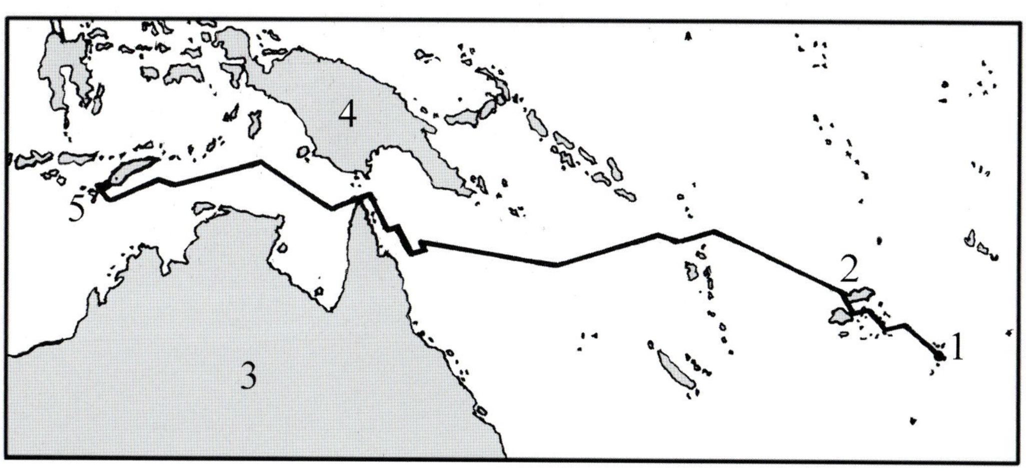

Map C: Bligh's Open Boat Voyage.

1. Mutiny site: 28 April 1789. Bligh and 18 of his followers are put into a 23ft launch.
2. Fiji.
3. Australia.
4. New Guinea.
5. Coupang: 14 June 1789. Bounty's launch arrives at Coupang.

14 June: A makeshift Union Jack is hoisted and the launch sailed into Cupang harbour. The mutiny is reported to Dutch authorities.

October 16: Bligh sails for Europe. Hall, Elphinstone, Linkletter, Lamb, Ledward, and Nelson all died of disease in Batavia or on their own journeys home.

1790

14 March: Bligh lands at Portsmouth and is fêted as a national hero. He was honourably acquitted for his loss of *Bounty* and promoted to Post Captain.

7 November: The 24-gun Sixth Rate frigate *Pandora* under Captain Edward Edwards was dispatched to return the mutineers to England.

1791

23 March: *Pandora* arrived at Otaheite and all fourteen surviving *Bounty* men surrendered or were captured. Edwards did not differentiate between mutineers and those held against their will and everyone was incarcerated in a prison cell erected on the quarterdeck dubbed 'Pandora's Box'.

10 May: *Pandora* departs Otaheite in search of the *Bounty*. McIntosh's schooner *Resolution* was confiscated and renamed *Matavy* to serve as *Pandora*'s tender.

Model of *Bounty*'s Launch, built by Bob Lightley, one of the finest ship modellers of his generation. The colour scheme of this model is speculative. *(© National Maritime Museum, London D3921_1)*

Lieutenant Bligh and his loyal crew are hospitably received by the Governor of Timor. *(© National Maritime Museum, London PAF7979)*

Searching until August with little success, they turned west and headed for the Dutch East Indies and home.

29 August: *Pandora* runs aground on the Great Barrier Reef and is abandoned. She sank before all the prisoners could be released – Stewart, Hillbrant, Skinner and Sumner drowned along with 31 of *Pandora*'s crew. *Pandora*'s boats were launched and sailed for Cupang on an open boat journey that largely followed Bligh's course of two years earlier. The prisoners were kept bound hand and foot until they all reached Cupang on 17 September. Edwards arranged for passage to Batavia and the prisoners were transported to Cape Town.

1792

5 April: The *Bounty*'s men embarked for England on the Fifth Rate *Gorgon*, arriving at Portsmouth on 19 June and transferred to the guard ship *Hector* to await trial.

Bligh, who had been given command of the sloop *Providence* for a second breadfruit expedition, had left England in August 1791, and successfully completed the mission.

12 September: The court-martial presided over by Vice-Admiral Lord Hood was held onboard the Second Rate *Duke*. Coleman, McIntosh, Norman and Byrne who maintained their innocence by voluntarily surrendering to *Pandora* were acquitted.

18 September: The remaining defendants were found guilty of mutiny and were sentenced to death by hanging, with recommendations of mercy for Heywood and Morrison.

26 October: Heywood and Morrison received pardons from George III and were released. Muspratt won a stay of execution by filing a petition protesting that court-martial rules had prevented his calling Norman and Byrne as witnesses in his defence.

18 October: Burkett, Ellison and Millward were hanged from the yardarm of the Third Rate *Brunswick* in Portsmouth dock. Muspratt was pardoned and freed on 11 February 1793.

In an ironic twist, breadfruit never became the staple food for the slaves and was used largely as animal feed. The Lobby's aim of sidetracking the abolition of slavery failed when on 15 March 1807, George III signed into law the *Act for the Abolition of the Slave Trade*, banning slavery in the British Empire, marking the beginning of the end of the most heinous practice in human history.

Model Products

There are *no* contemporary illustrations of the *Bounty* but despite this over 30 kits have been released in wood or plastic beginning in the 1950s. The best references are C Knight's 1936 article 'HM Armed Vessel *Bounty*' published in *The Mariner's Mirror* that was greatly expanded in 1986 with John McKay's book *The Armed Transport Bounty*. McKay notes that his depiction of *Bounty*'s decoration is conjecture and based on the sailing replicas that were the source of inspiration for kit manufacturers, rendering virtually every model of *Bounty* built a model of one of the motion picture replicas.

With a few exceptions in plastic, virtually every kit of the *Bounty* is a wooden ship model. A helpful resource on building wooden kits is this author's *Ship Models from the Age of Sail: Building and Enhancing Commercial Kits*, and for rigging, Lennarth Petersen's step-by-step *Rigging Period Ship Models* is highly recommended, as is membership in the Nautical Research Guild, whose website and publications feature extensive reviews and build logs for hundreds of wooden kits making it an invaluable resource.

CALDERCRAFT HERITAGE SERIES **HMAV *BOUNTY*** 1:64 Scale

The Caldercraft *Bounty* is of double plank-on-bulkhead construction where the hull is built by gluing shaped bulkheads to a false keel (a wooden profile of the hull) cut from plywood to form a framework that is planked over twice with wood strips. The first planking is typically in a softwood like lime to form the basic hull carcass, followed by a second layer in a thin hardwood like walnut. The advantage of double planking is that the first layer allows you to shape the hull and correct any dips or bumps before planking over a second time with the finish wood. A thin ply subdeck is glued to the hull and planked over with strips of Tanganyika wood.

The hull and fittings, from the locker over the rudder head, tiller, capstan, windlass to the figurehead, accords well with McKay's drawings. This kit also reproduces many of the details unique to *Bounty*, such as deck hatches incorporating ventilation ports. The masts are fitted with gratten tops, but the gratings are represented by a series of round holes drilled through the plywood part. The kit's components are precisely cut and once the whiskers of wood are cleaned away they fit together well.

Detail pieces such as gun carriages, headrails, rails, binnacle, and mast top supports cut from walnut.

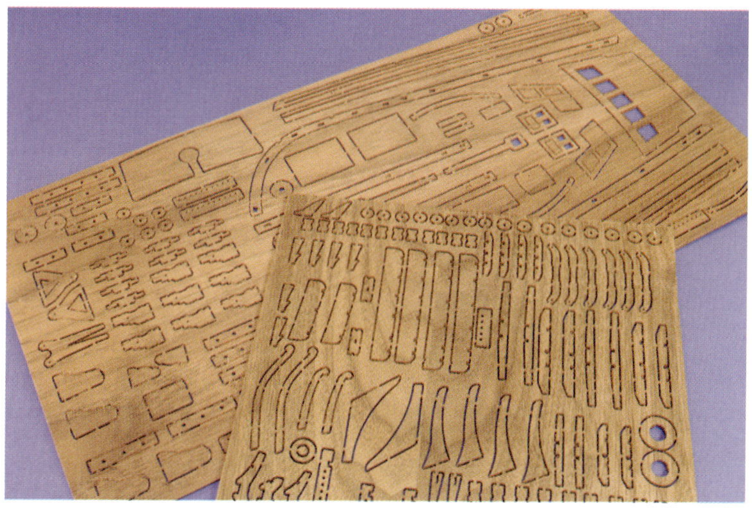

Most of the detail fittings are cleanly cast in white metal. The quarter badges and stern decoration replicates McKay's drawings exactly. The swivel and 4-pounder carriage gun barrels are turned brass items. The swivel guns fit into fold up photoetched brass cradles whereas the 4-pounder guns fit to carriages assembled from walnut. The brass etched sheet contains the ironwork required for rigging and provide deadeye strops, chains and chain plates, frames for the *stern lights* (windows), and a decorative transom piece sporting the name *Bounty* rendered in an unrealistic modern font.

The hull's bottom is sheathed with dimpled copper plates. The dimples represent nail heads and are best hammered down for a more realistic effect. A bag of fine brass eyelets for eyebolts and jackstays are included in addition to hundreds of miniature brads to tack the first planking to the hull bulkheads.

The model's construction is described in a booklet of written instruction supplemented with a large set of scale and perspective drawings. The rigging plan is simplified with the ship in *harbour rig* (no sails). The over-scale belaying pins are turned from walnut. The rigging blocks are generic rectangles of walnut (common to many wooden kits) with drilled holes – not always accurately – to represent the sheaves through which the rigging lines are *rove* (threaded through). A good selection of linen thread dyed tan and black is provided for the rigging line. The black thread represents the tarred *standing rigging* (that forms part of the structure of the ship and

Components for the ship's boats and gratten tops.

Above: Photoetched brass sheet providing strops, chains, plates for rigging, window frames, oars, hooks and grapnel anchors among other items.

Above, right: Turned brass gun barrels.

Right: Dimpled copper plate to sheath the hull bottom.

keeps the masts in place), and tan for the untarred lines of the *running rigging* (the lines that are used to work the sails and yards). The deadeyes are turned from walnut and are authentically stropped and attached to the hull with the chains and plates from the etch brass sheet. No flags are included.

Two ship's boats are included – the launch and the cutter (erroneously called the 'jolly boat') that are small plank-on-bulkhead kits. After planking, the tops of the bulkheads are removed and the boat's floors, thwarts and interior details made up from wood strip. What is provided for the boats is very basic, and for a realistic effect, details such as frames, gratings, and knees can be fashioned from scrap wood and added. The boats should be filled with oars, casks and a mast and some furled sailcloth for realism.

There are two challenges to be aware of. Firstly, the position of the mizzen mast, bitts, fittings, binnacle and wheel indicated on the kit's false deck are too close together. The modeller is well advised to check their location with McKay's drawings and space accordingly. Second, the anchor cables as drawn on the plan are shown wrapped around the windlass incorrectly. The cables should lead over the *top* of the barrel (and therefore run aft off the top as well) and there should be only three turns. This is clearly shown in John Harland's *Seamanship in the Age of Sail* on p263. The sketches in this reference show that lead of the cable can lie inboard or outboard depending on what lines up best with the fore hatch through which the cable ends would have been run to the cable tier. At any rate, when at sea, the anchor cables are *unbent* (taken off the anchors) and stored in the cable tier for protection. The anchors themselves are lashed to hang off the ship's side with the flukes hung over the rail.

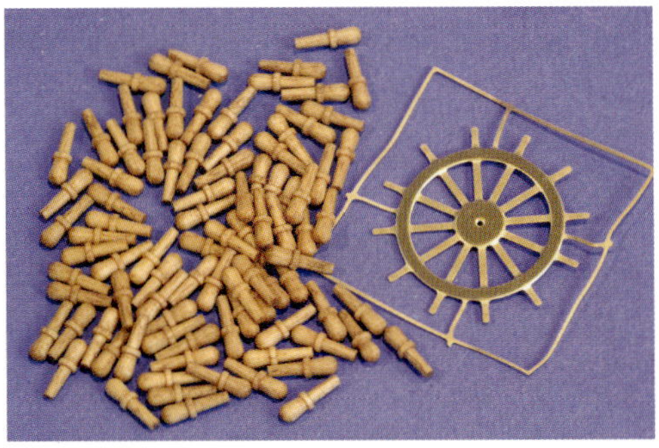

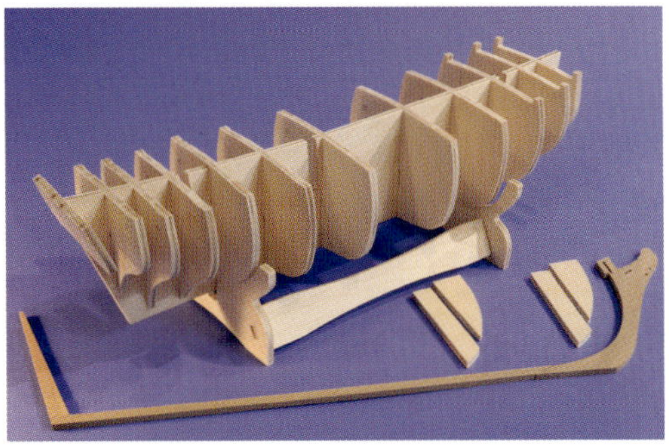

Above, right: Unrealistic generic rigging blocks.

Right: Etched brass ship's wheel and over-scale turned wooden belaying pins.

Far right: Bulkheads attached to the false keel.

DEAGOSTINI 'ADMIRALTY STYLE' *BOUNTY* 1:48 Scale

The term 'Admiralty model' is a modern term for Navy Board models whose hulls were unplanked to provide the Lords of the Admiralty a three-dimensional representation of the sweep and shape of the hull. These exposed frames are not to be

Far left: Each monthly part arrives in an oversized box.

Left: Each month's selection of parts is wrapped in plastic. Just a few months' collection is shown here.

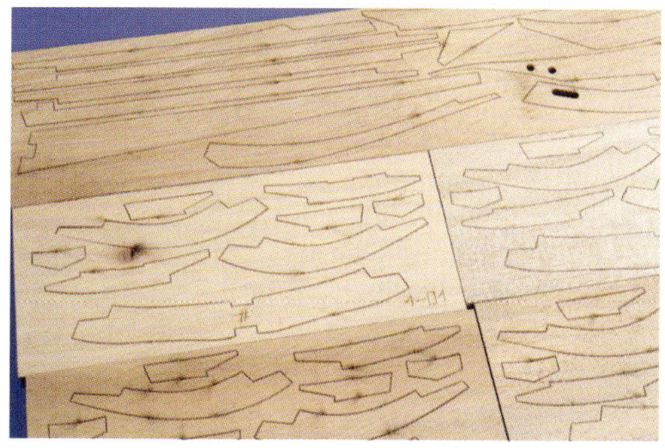

Left: Typical selection of laser cut components. At the top are the stem and stern pieces, and below that four of the nearly one hundred (!) sheets containing the frame components.

confused with bulkheads typical of wooden kits. Rather, these frames were stylised representations of the actual ship's frames, as best explained in John Franklin's *Navy Board Ship Models 1650-1750*. Today, 'Admiralty style' now refers to *any* model that is built with exposed frames, whether the disposition of the frame and ship's timbers are accurate reflections of the time or are stylised versions. This form of model has long been considered the epitome of model shipwrightry because each timber must be individually crafted by hand from fine hardwoods. These models are often built from copies of the original ship's draughts to *loft* (plot out and draw) each frame and timber for the model, and all decorations are hand carved.

Today's technology has put this type model within reach of all modellers by lofting the thousands of timbers with 3D modelling techniques. The ship's timbers are cut using lasers with figureheads and decorative sculptures carved using 3D routers. Cast brass and resin parts, often supplemented with 3D printed parts provide high-definition fittings, with etched brass components providing rudder hinges, brackets, chains and strops. In many kits, even the rigging blocks are milled in wood using 3D designs giving each block the proper shape, scored for strops, and realistic sheave detail.

DeAgostini have created a framed model of *Bounty* but be aware that no such model of *Bounty* ever existed. The McKay drawings appear to be the basis for the structural timbers but the model differs in a number of details. For example, the number of frames is *half* of what was used to frame *Bounty*'s hull so the model's interior and breadfruit pot racks can be more readily seen. The kit parts are split into 12 'chapters' that are purchased by monthly subscription. Each chapter is structured in a way to provide you with enough varied tasks to keep you from getting bored until the next month's instalment arrives. For example, in the first few sets you assemble and paint a swivel gun, build the keel parts, and make a start on the elaborate building jig that holds everything together.

No scale plans or printed instructions of the ship are provided and must be downloaded from DeAgostini's website. The lack of scale plans is a problem because there is no real way of checking if your construction is accurate. For example, the fit of the model into the building jig is quite slack and there is no way of knowing if this slackness is deliberate to facilitate taking the model in and out of the jig or if your construction of the hull is somehow undersized. Similarly, with the deck beams: some of them are cut to approximately the correct length being just a little over-long, but some are much too long. It makes you wonder if the parts were cut correctly or your hull is too narrow. At the bow some of the beam notches are clearly in the wrong place, and there is no way to check where they should be except *in situ*.

The downloadable instructions are very well illustrated with lengthy step-by-step illustrations, but working from them on a computer or tablet at the workbench can be annoying. Printing off the hundreds of

Right: The construction jig to hold the hull frames in place.

pages – many of which unnecessarily show the same assembly instructions for a single timber from several perspectives – is a waste of paper. A set of scale drawings would have been infinitely preferable.

The kit was designed somewhere in continental Europe and the English instructions are a machine translation with the usual awkwardness. DeAgostini did not use the proper names of the ship components. For example, the frame parts are just called that – no attempt to call out which item was a floor, futtock, top timber, or chock. The instructions are almost entirely pictorial and give vague advice like 'use a sanding sponge to gently alter the shape of the part', rendering the model best for those who are familiar of actual ship construction and have woodworking experience.

The kit parts are laser cut from an unidentified semi-hardwood with some noticeable grain and flecking. Laser cut charring has been kept to a minimum with almost no scorching on the face of the parts. The numbers for each component are not etched onto the carrier sheets, requiring tedious back and forth referral to the parts maps that require endless scrolls on the tablet to locate.

The model is built into a slip made from medium density fibreboard. It features laser etched markings to ensure that the correct frames are put into the right location. The frames themselves are constructed from several parts into a single layer. These parts are built over 1:1 size templates that are downloaded and printed onto a sheet of paper. Colour coded templates are provided to help you sand in the correct bevels.

The launch is a true plank-on-frame kit in itself. It is built over a jig and much like a real boat. Laser cut frames are placed into the jig and the keel parts fit into the notches. The hull is planked and instructions provide the dimensions and tapers of each plank to fit the hull contours.

Ship's fittings such as the capstan, pumps, and windlass are assembled out of several wooden parts and detailed with neatly cast zinc or photoetched brass components. The resin figurehead is poorly cast and will need considerable clean up prior to painting. Photoetch sheets of brass provide deadeye strops and chain plates, the ship's lanterns, and many straps. The brass is very thick and the parts will need to be annealed before bending. Decorative items such as the stern detail and quarter badges are laser etched wood. Six representative breadfruit pot racks are provided for the stern cabin.

As befits an 'Admiralty Style' model, only stub masts are provided. The woodwork is left almost entirely unfinished like actual Navy Board models, but some parts of the model can be painted for effect. It is worth studying actual Navy Board models held by the National Maritime Museum at Greenwich, which will give ideas as to how to finish the wood and what aspects can be painted to better emulate this style of model.

This model will provide a deep appreciation of how sailing ships were really built especially if you take the time to look up the name of each timber in a book like Peter Goodwin's *The Construction and Fitting of the Sailing Man of War 1650-1850.* You'll see firsthand how each timber supports the next, and that as the ship comes together how strong and stout they were. Remember to photograph the model under construction so there is a record of the framework before it gets covered in plank – although you can leave off or add as much planking as you like.

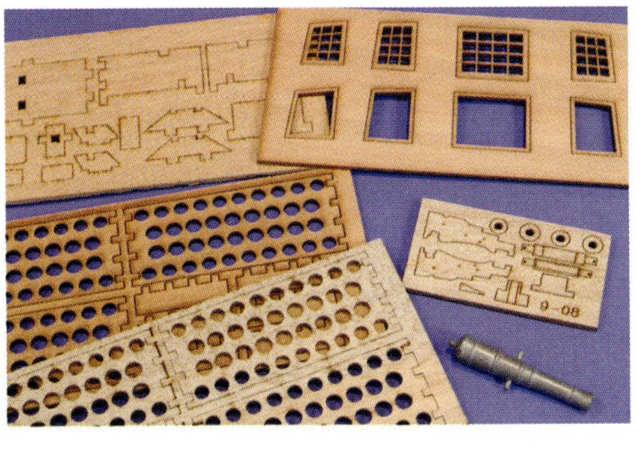

Right: Laser cut components for the pot racks (bottom), gun carriage (right), locker (top left) and lights (top right). The light frames are a different shape and style from all other models and drawings extant.

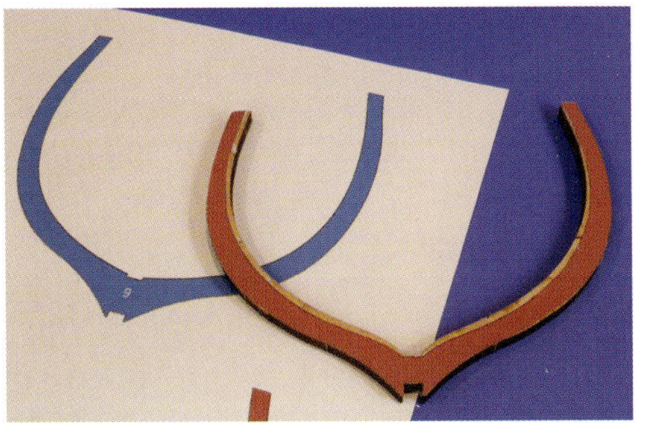

Far right: An assembled frame with a bevelling guide (red) attached to one side. The blue one is pasted to the back of the frame and the exposed wood is carefully filed away to the proper bevel.

AIRFIX **HMS *BOUNTY*** 1:87 Scale

Ever since MGM's 1935 motion picture *Mutiny on the Bounty* starring Clark Gable and Charles Laughton, *Bounty* has been bestowed the title of *His Majesty's Ship*, but the use of *HMS* began around 1789 *after* she embarked on her mission.

The hull castings carry fine recessed hull planking lines and copper plating and the main kit parts accord well with Knight's 1936 general arrangement drawings. The

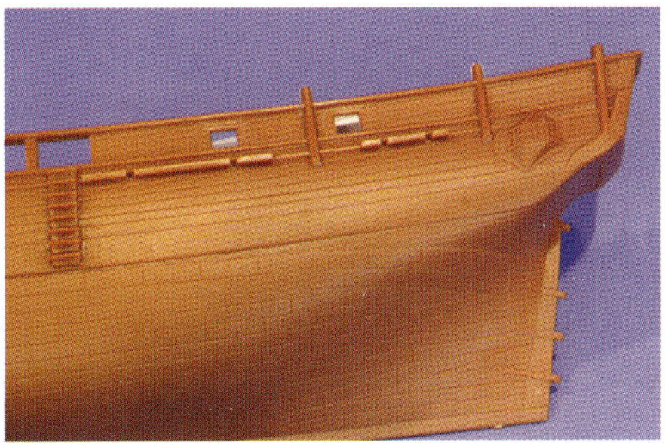

deck is a single piece with scale planking and all hatches and gratings moulded in place. The gratings are solid but the lath and slat detail is just about to scale and careful painting and highlighting will give a good impression. The gratings at the bow are provided as hatches with hinged lids, and any ventilation holes need to be drilled out. The deck fittings are finely moulded – and in some cases too fine. For example, the bitts and the windlass cheeks are very thin and would never have taken the strain of pulling ropes so they are best re-made from thicker sheet stock or square rod.

The kit includes hulls for *Bounty*'s original allocation of boats but the small jolly boat should be omitted. The kit includes a few figures to populate the deck – one presumably being Bligh (recognisable as Charles Laughton with a protruding belly no doubt due to his consumption of the stolen cheese!) and a few seamen swabbing the deck or carrying barrels of water and chests to provide a sense of scale.

Like all plastic sailing ships, the biggest challenge are the masts and yards which being thin moulded items are quite prone to bending under the strain of any rigging. The upper and lower masts are best replaced with items made from tapered hardwood dowel, and brass rod for the topgallants. The main and upper yards are easily tapered at both ends from hardwood dowel, and straight brass rod for the topgallant yards and stunsail booms would suffice. The bowsprit and jibboom are moulded as a single chunky unit and would be best re-made as two separate parts. The mast tops are solid and easily converted to gratten tops by sanding them flat and laying etched brass mesh over them.

Rigging a plastic model ship is another challenge. The kit provides the deadeyes, channels, and chain plates moulded as a single unit that are glued to the hull sides. They are acceptable when looking at the model from three feet away, but not upon close inspection. This is because the dead-

Above left: The hull casting carries fine and accurate detailing.

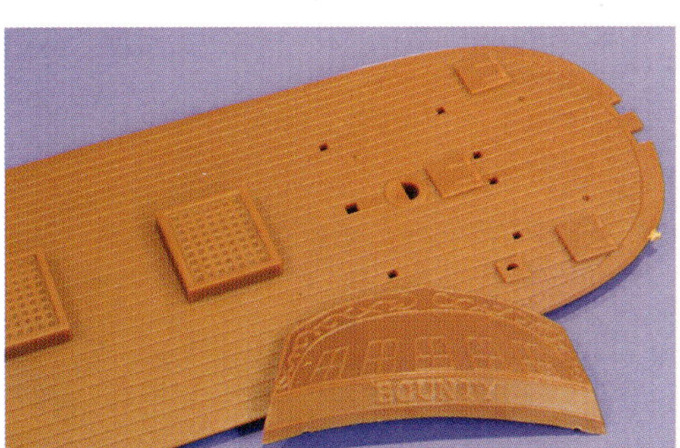

Far left: The deck and stern decoration is very well done. The ship's name is executed in a period font.

Left: The deadeye and chainplate details are provided as a single unit. The hull sides feature authentic detailing such as the side steps.

Right: Which one is a mutineer and which loyal to Bligh?

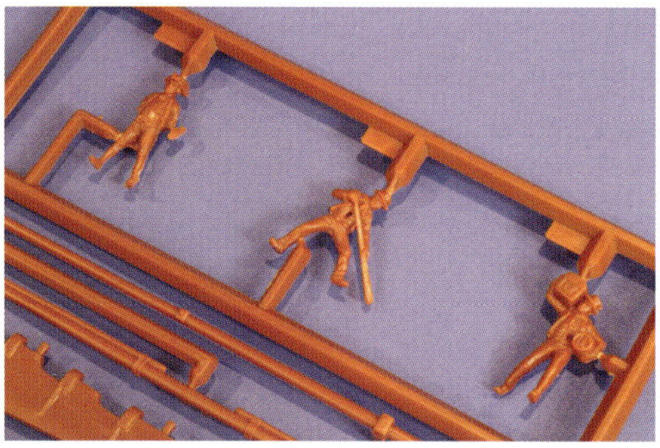

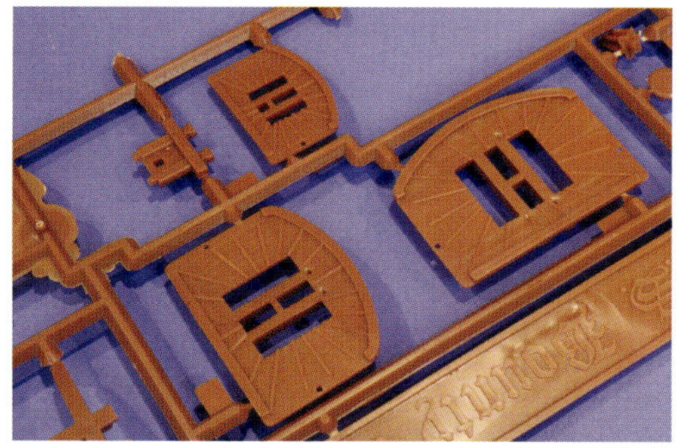

eyes are simple pierced rings so thread can be tied to them to form the shrouds. After attaching the thread, a blob of paint or glue to the front of each deadeye will give a more convincing deadeye. Alternatively, adapting miniature shipbuilding techniques that use punched card disks and wire as described in Phillip Reed's *Modelling Sailing Men of War* or *Period Ship Modelmaking* is ideal.

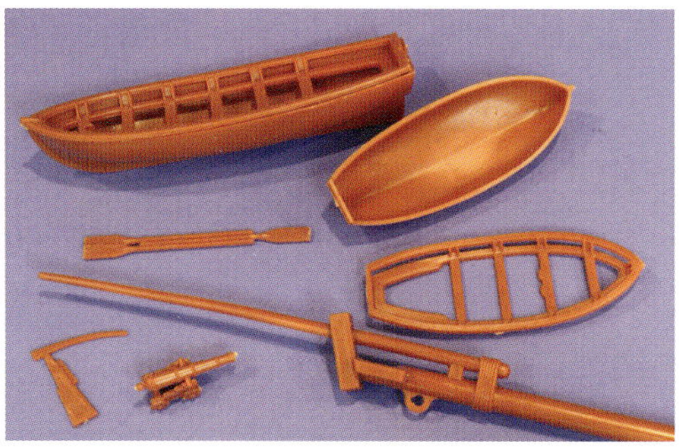

Right: The boats are complete with oars and rudder. The jibboom and bowsprit are chunky items.

A single spool of thick white thread is included for the rigging lines, and the rigging plan is quite simplified. A loom is provided to weave shrouds and ratlines as a single unit, but if used as instructed, the shrouds and ratlines would be erroneously the same weight: a heavier weight thread should be used for the shrouds and a light-weight thread for the ratlines. The loom does work in practice, but it is difficult to produce the correct spacing and *rake* (angle of the shrouds). Sails are thin vacform items that are fine for younger modellers but best omitted, leaving the ship in harbour rig.

HiSMODEL ACCESSORIES FOR THE AIRFIX **HMS *BOUNTY*** 1:87 Scale

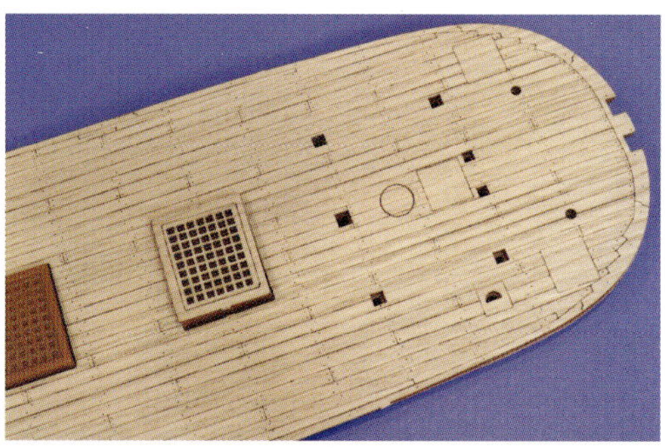

Right: The wood overlay fits the Airfix plastic deck precisely.

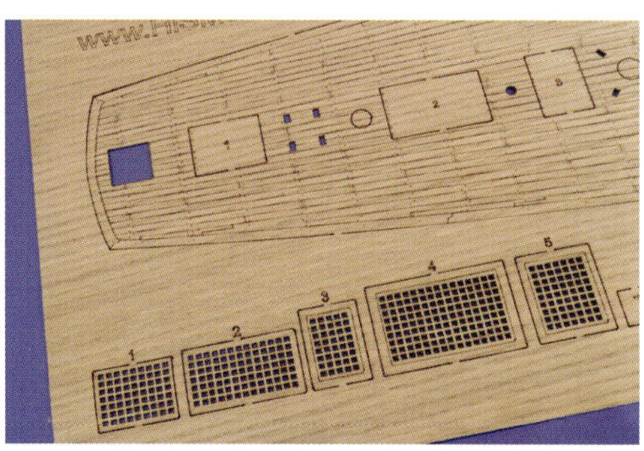

Right: Laser cut gratings for the deck hatches.

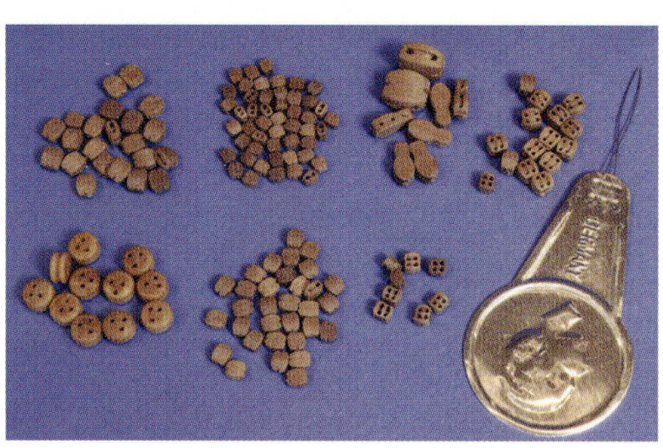

Right: Even in this small scale, the rigging blocks are authentic and perfectly shaped with sheave detail.

HiSModel from the Czech Republic produces a huge range of accessories for plastic and wooden sailing ship kits. For the Airfix kit they have created several items to build a super-detailed model. The first item is a laser cut deck overlay in oak veneer that features etched planking with the proper shift of butts and trennel detail. At the bow and stern where the ship's hull narrows, the planks are properly nibbed into the margin plank. The overlay fits precisely over the plastic deck, and also includes scale wooden gratings to fit into or over the hatch coamings.

A real star item is a set of CNC cut and sewn sails from a very fine weave cloth with stitched detail. The fabric is pre-dyed a realistic beige colour and every sail that *Bounty* carried is provided. With these sails (or even without them), you will want to rig your model just like the actual ship and HiS offers a *Bounty*-specific set of rigging wooden blocks, deadeyes, and threads. In

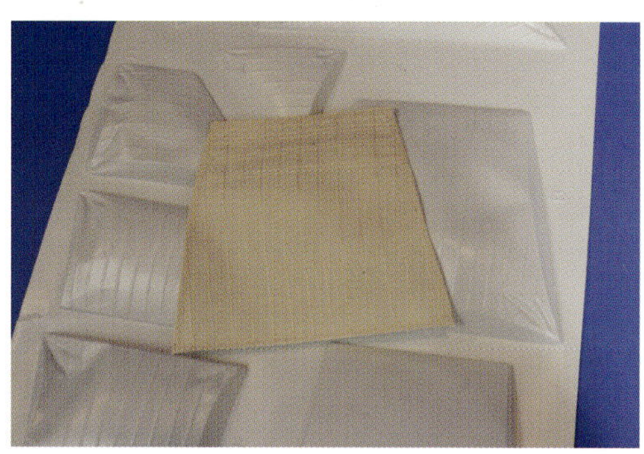

Far right: The contrast between the cloth and vacformed kit sails is dramatic.

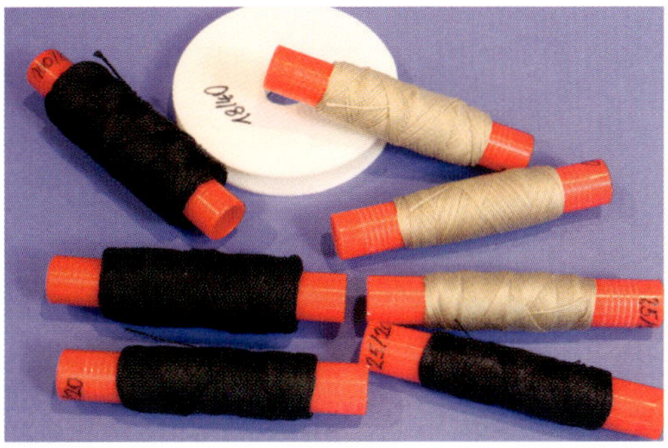

this small scale, many of the double and single blocks are tiny but perfectly milled to the correct oval shape with sheave detail. All of the holes are drilled through and are designed to be properly stropped and fixed to the mast and yards the same way master riggers did to provide the ultimate in detail. The rigging line set comprises five thicknesses of black and four of tan thread in scale to the model. To help the modeller choose the right block and line, HiS Models has produced a series of guides based on the McKay drawings showing where and when to use a specific size and shape block and thread. Their guides are very comprehensive and will ensure your model will have the most accurate scale rigging possible.

The 4-pounder guns are available as items with turned brass barrels and laser cut carriages. Several pre-blackened fittings such as eyebolts, hooks, and an extensive parrel set provides further essential detail. Cloth flags and pennants are available to finish off your model. Although the sets shown here are designed to complement the Airfix kit, HiSModel can provide rigging sets or individual items for *Bounty* models in other scales and materials. Simply let them know what ship you are building, the scale, and manufacturer and they will provide you a list of what accessories they can provide. Highly recommended.

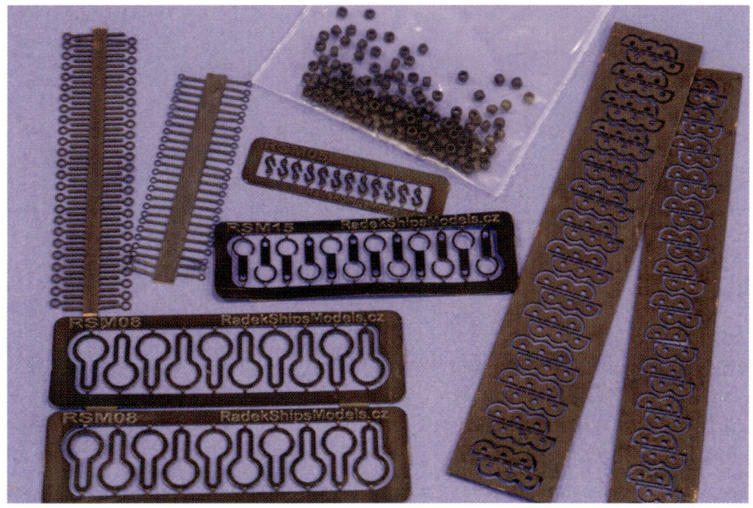

Above: Pre-blackened etched metal parts – deadeye strops, hooks, eyebolts, and parrel assembly.

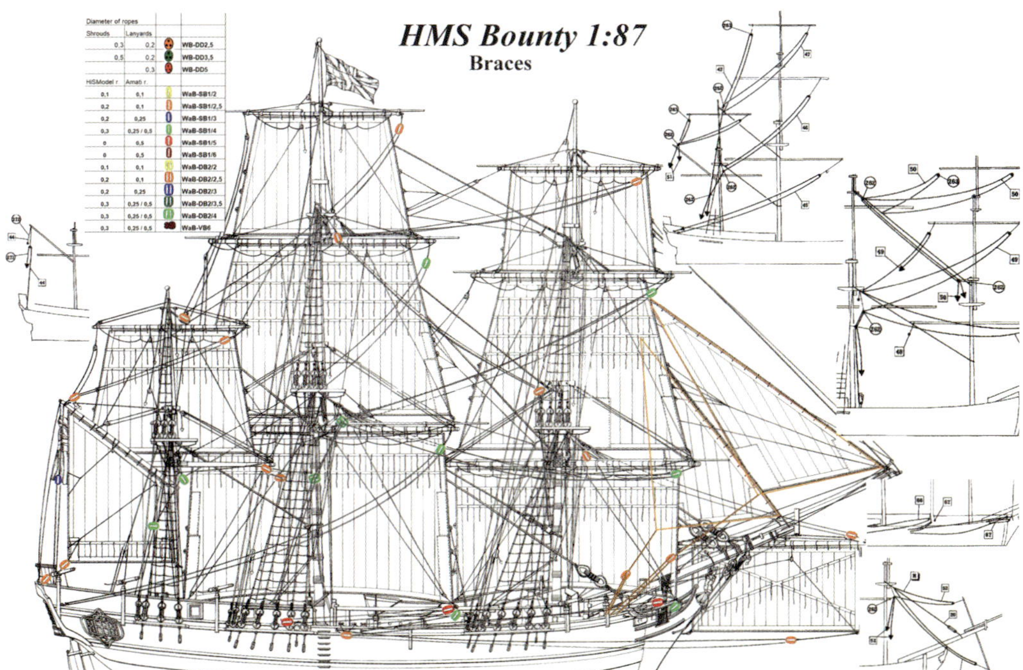

Left: An example of the extensive diagrams for the rigging sets. All of the diagrams are found on the HiSModel website. Rigging sets can be adapted for use with a *Bounty* model of any scale by request.

AMATI HMS *BOUNTY* 1:60 Scale

The box lid describes the kit as of 'museum quality', and trumpets its 'historical accuracy' and 'fidelity of scale'. Is this true? Opening the box reveals trays of very well produced fittings in brass, wood, white metal, laser cut plywood, printed sailcloth, photoetched brass, and beautifully drawn instructions. The metal parts are cleanly cast and the wooden parts are precisely laser cut. The kit is planked twice

Left: Generic guns and fittings.

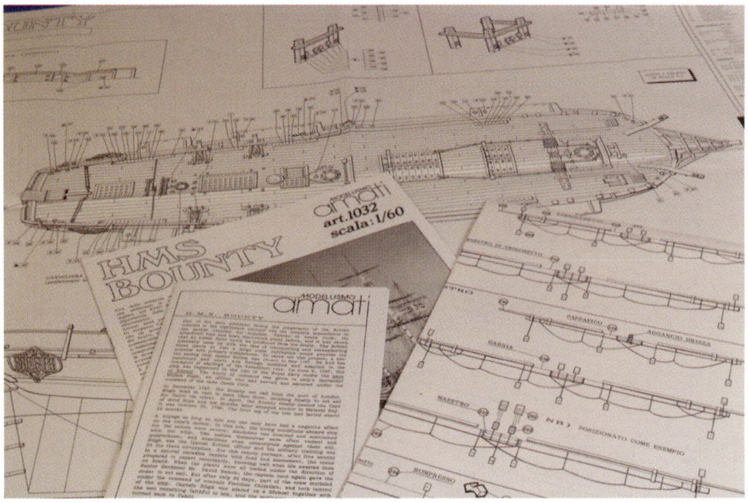

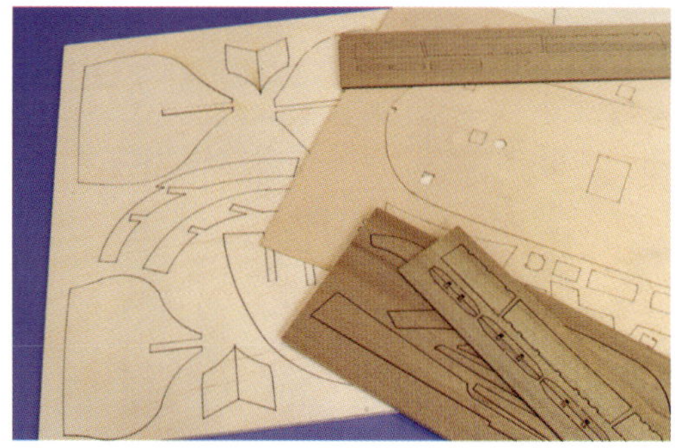

Above: Beautifully drawn plans and comprehensive written instruction book.

Right: Etched brass stern decoration and white metal quarter galleries and figurehead.

in walnut strip clear of knots and blemishes. The mast and yard material is fine grained walnut dowel. All high-quality parts indeed.

However, closer examination shows a model built out of the box will only produce a broad representation of *Bounty* because Amati has relied heavily on generic items from other kits in their range giving the model a toy-like appearance. For example, the over-large gun carriages carry prominent timber and bolt details. The cast gun barrels sport an eye-catching but imaginary continental style coat of arms. The lanterns appear to come from an earlier period with beautifully turned brass tops but do not resemble the pattern shown on any *Bounty* drawings, and come with dramatic red plastic inserts for the glass. The lanterns' sides also carry delicate filigreed cut outs that are wholly inaccurate; and important details such as copper plates for the hull are absent. Only a single ship's boat, presumably the launch, is provided as a generic white metal casting with walnut strips to simulate boat frames, thwarts and other details. The boat sits on chocks of an imaginary design and the painting instructions are quite fanciful – going beyond the Hollywood-inspired replica ships by adding white lower masts. The mast tops are solid. On a positive note,

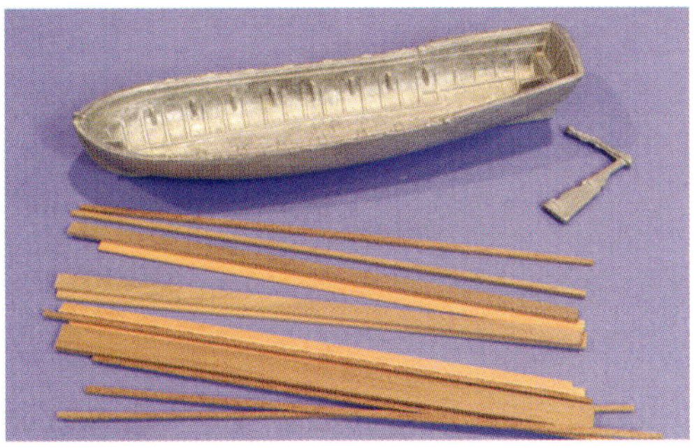

some cast metal figures are provided that do give a sense of scale.

Spools of good quality polyester-cotton threads and hundreds of generic blocks are provided for the rigging. Turned walnut deadeyes are authentically stropped into shaped wire strops that are linked to wire loops that attach to brass strip chain plates on the hull. Cloth sails are printed in dark brown ink on a fine off-white colour cloth. All of the cloth bolts, reinforcing patches, and seams details are printed only on one side of the cloth, so they will have to be traced on the backside with a dark brown pencil or a pen made for fabric. The rigging plan is simplified and a sheet of vinyl flags, carrying a Grand Union, Blue, White, and Red Ensign are provided.

In terms of 'historical accuracy' and 'fidelity of scale', this kit does not live up to its billing, but happily the basic hull shape and general layout accords with references, enabling the kit to be reworked into an authentic replica.

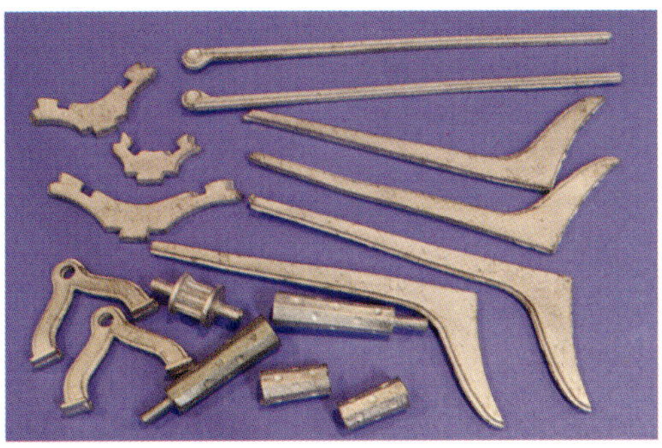

Above: Generic fittings in white metal, brass, plastic, and bronzed white metal.

Above, left: Generic white metal launch with wooden strips for detailing.

Left: Headrail, capstan, and ship's wheel brackets finely cast in white metal.

OCCRE ***BOUNTY*** 1:40 Scale

Occre of Spain offers an entirely unique large scale *cutaway* model of *Bounty* designed to show off a complete interior outfitted with cabins, holds, bread-

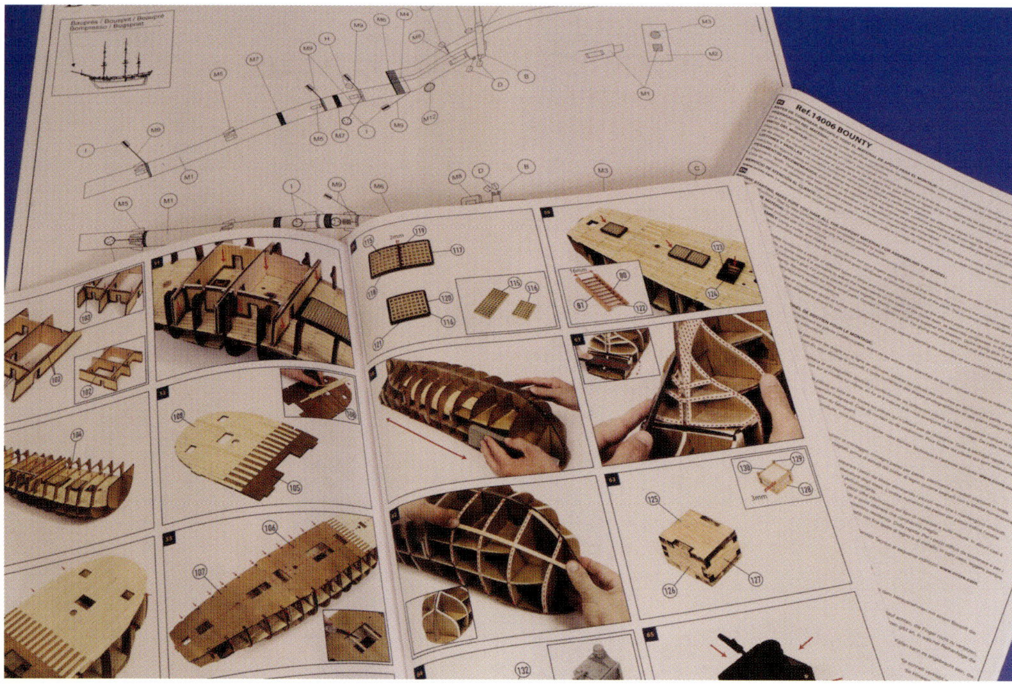

Right: Fully detailed instructions guide you through assembly. This booklet is supplemented by beautifully drawn 1:1 scale plans.

fruit pots and other fittings that can be viewed unobtrusively through the hull.

Opening the well packed box reveals a lot of thought went into this kit. The instructions consist of a full colour illustrated step-by-step booklet supported with written instructions in several languages.

Below: A selection of laser cut hull and launch components in ply.

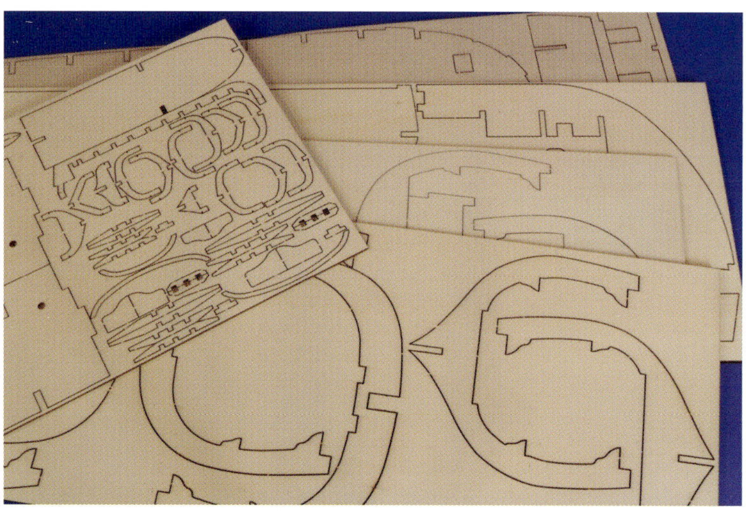

These instructions go beyond showing how the model should be put together, but also teaches basic wooden shipbuilding techniques such as how to taper planks, bevelling, simulating caulking and trennel detail. A set of 1:1 scale printed plans of the hull assembly, the masts, yards, and where rigging blocks, jackstays and eyebolts are to be seized leaves no doubts. This is probably one of the best sets of instructions available for any wooden kit.

The hull is double planked and the wooden components are neatly laser cut from plywood and walnut sheet. Hull planks are blemish-free strips of sycamore and sapele. Being a cutaway model, the bulkheads are spaced widely apart and care will have to be taken with the first planking to minimise any sagging between them. For the interior spaces, representative numbers of breadfruit pots, barrels, and other stowage is provided. The enterprising modeller can fill the ship to the brim with ballast, with more plant pots and racks, bunks, hammocks, sacks of corn, biscuit, tools, casks and even some livestock. Researching what a ship carried at sea makes for some fascinating modelling. Stephen Beisty's *Cross Sections: Man of War* is a fun way to see how wooden ships were stowed and provisioned.

There is a full set of sewn sails on fine

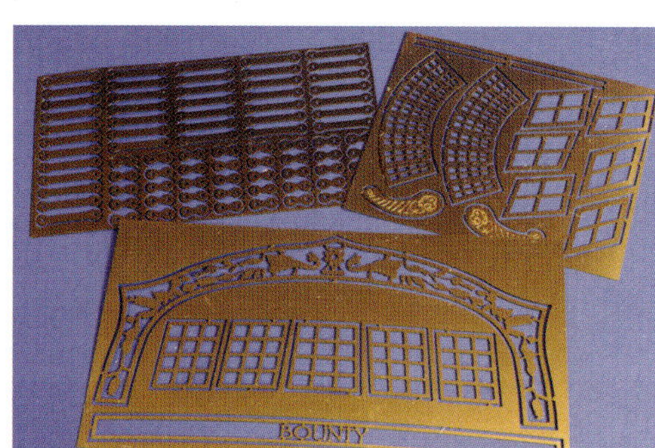

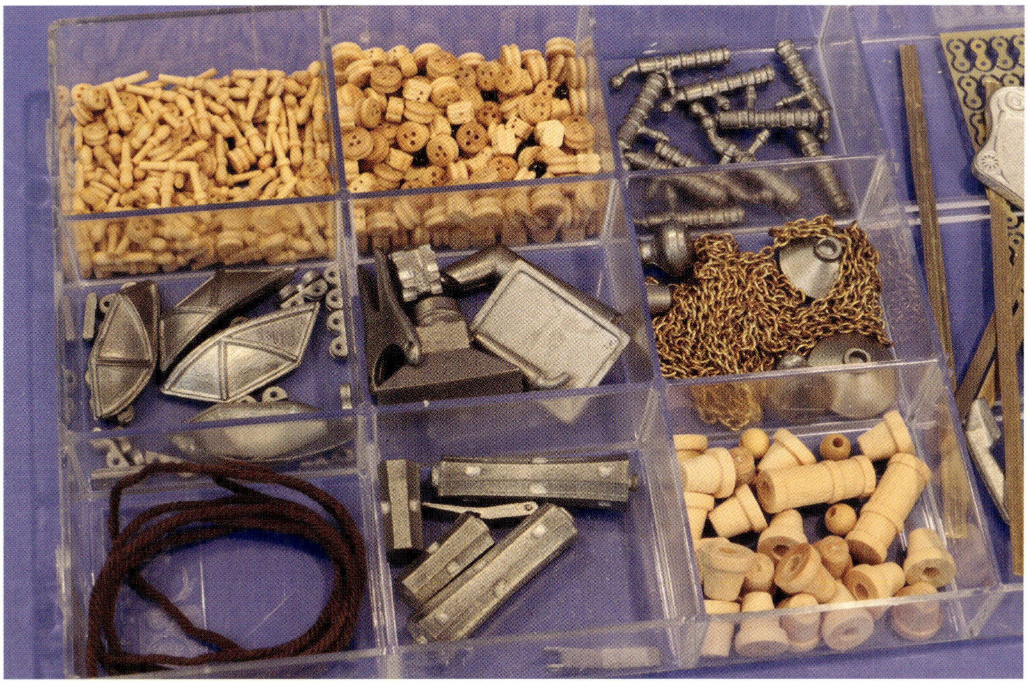

Left: A selection of well detailed fittings. Only the swivel guns are out of scale and shape. Breadfruit pots are turned wood.

weave tan coloured fabric complete with 22 spools of thread for rigging in dark brown and light tan. The rigging blocks are generic but given the large scale of the model they are easily sanded to a more realistic shape, or replaced with accurate milled items. The turned deadeyes are stropped by looping each with wire and twisting the tails tight which is a convenient modelling expedient, but proper strops and chains are best scratch built from wire. Authentic gratten tops for the masts are assembled from notched wooden strips. An illustrated rigging plan is included and shows the modeller the *direction* lines are to be rove through blocks minimising guesswork and fouled lines. Like most kits, the rigging plan is simplified in places and consulting authoritative reference is recommended.

The kit's fittings are a mix of generic items and bespoke to *Bounty*. Generic items include the swivel guns which resemble those from the sixteenth century. The 4-pounder guns feature a large round muzzle swell characteristic of early French guns but can be reshaped with a file. The capstan is a little undernourished, and the flags are a White Ensign and several St George's Crosses. The bespoke items include etched brass stern lights, transom decoration, and quarter badge base. *Bounty*'s launch is included as full plank on frame mini kit. Breadfruit pots are included, and young plants made out of supplied wire. Using model railway scenery techniques, you could make full scale leafy, healthy breadfruit plants to fill the garden in the great cabin.

The kit adopts a 'paint with wood' finish that uses different species of wood to produce a decorative model. For example, the interior is built of light coloured wood with trim in walnut; and the pump barrels are in a light coloured dowel detailed with black bands and walnut handles. Happily, the model is accurate in form, so if an authentic finish is preferred, the wood can be simply painted. For example, the interiors might be whitewashed as was the practice of the day, the brass fittings properly blackened with paint or jeweller's patina solution, the pumps painted over brown or black, and the launch and oars fully painted in ochre and white. Like most kits of *Bounty*, the hull bottom is erroneously painted white on this model and for an authentic finish will have to be coppered.

Above: A full suit of sails in tan fabric with slightly darker tan stitching.

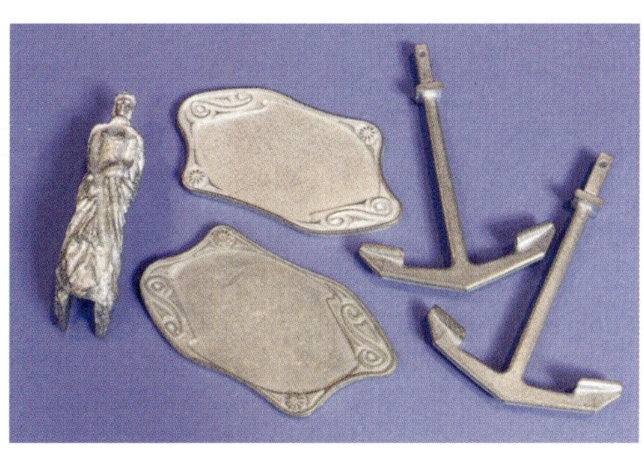

Left: White metal figurehead and quarter badge bases. The quarter badges are constructed from cast white metal and brass etched components for a fully hollow appearance. The anchors provided are definitely not naval pattern, but it is not clear if the anchors carried by *Bounty* were those purchased with *Bethia*, or replaced with those of a naval pattern. The kit supplied anchors may be the correct merchant pattern.

The kit does not provide a stand to hold the model whilst building or for display but Occre offers a building slip to hold your hull that can be used for any other model. For display, they offer a very nice stand that when finished is an excellent complement to the model.

Built out of the box the Occre kit builds into an eye-catching model of *Bounty*. It is quite a remarkable package not only for its competitive price but because of its huge potential in the hands of a modeller who can really expand on fitting out the interior, adding more of the ship's structure such as beams and knees, perhaps lighting, and some crew members from wargame figure ranges (28mm to 30mm figure size). Occre's model is quite a display piece that will be the attraction in any room however it is built. Highly recommended.

MAMOLI **HMS *BOUNTY*** 1:135, 1:100, 1:64 Scales

Above: Kit boxes for all three Mamoli *Bounty* kits.

The venerable Italian brand Mamoli was founded by Luigi Volonte over 40 years ago and is responsible for popularising wooden model shipbuilding around the world. Dusek Model Ships recently acquired the brand and has been upgrading kits with high quality materials and modern production standards but set out to retain the charming character of the original Mamoli designs. They have also introduced new kits into the range to suit the different needs of modelmakers today.

MINI MAMOLI **HMS *BOUNTY*** 1:135 Scale

This miniature version of the *Bounty* is a recent design by Daniel Dusek aimed at modellers new to sailing ship subjects. However, this kit stands apart from all other beginner kits due to its scale fidelity. The kit's centrepiece is a perfectly shaped 3D printed solid hull that has been impregnated with wood particles. As a result, the hull resembles a wooden machine-carved piece that can be sanded and worked the same way. Out of the box, the hull only needs some smoothing before fixing a laser cut wooden deck, keel, stem and sternposts into slots ready moulded into the hull. The hull features a ready-made rebate along the deck edge to neatly accommodate laser cut wooden bulwark pieces. The laser cutting is very clean and features laser etched details such as deck planking detail and stern decorations, lights and the quarter badge details. Fittings such as the capstan, guns, ship's wheel, and figurehead are cleanly cast in white metal. The kit instructions are easy to follow so the neophyte modeller is left with little doubt about how to do things.

The masts and yards are made from birch dowels, and a single spool of thread is provided for rigging. The mast tops are laser cut but solid. In this small scale the gratten tops can be drawn in with pencil or some etched brass square mesh can be glued to the plain tops to represent them. Given that

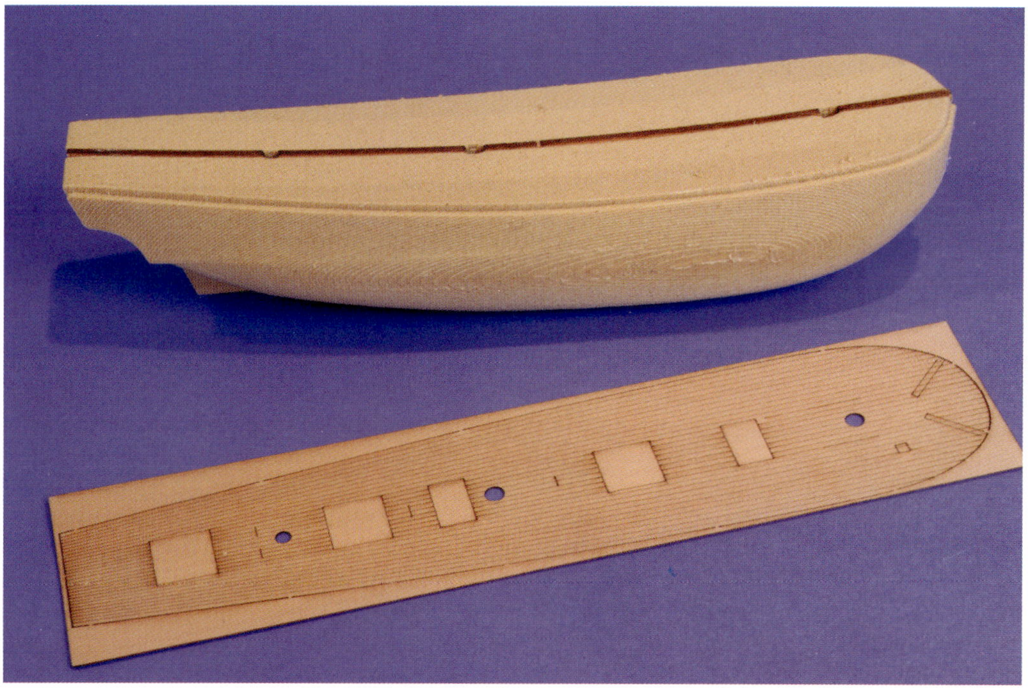

Left: The 3D printed wooden hull and laser cut and detailed deck.

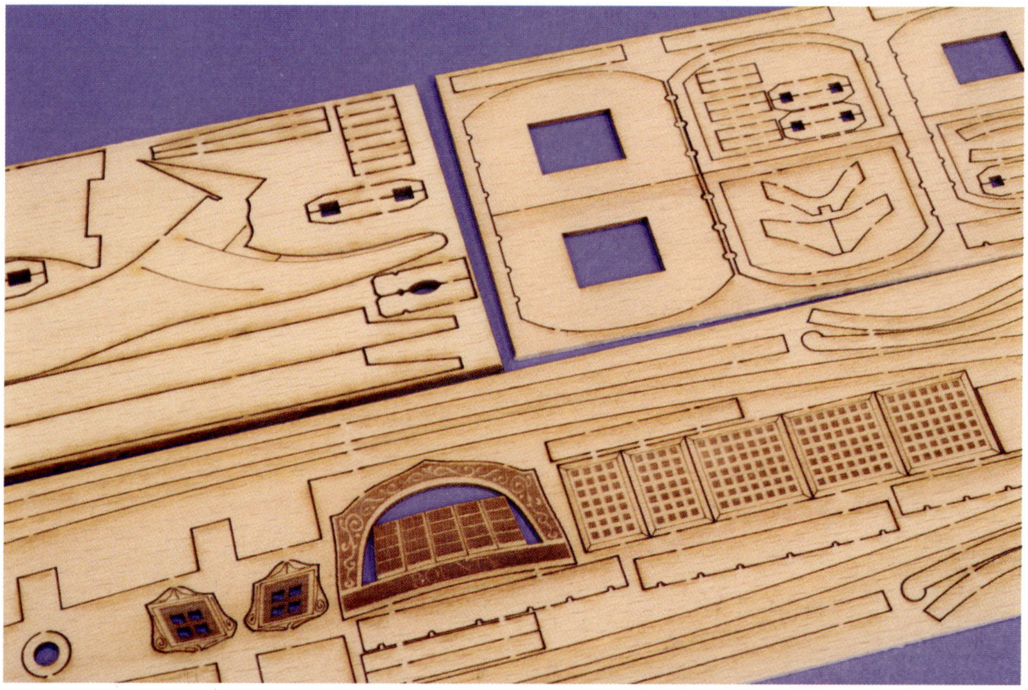

Left: Laser cut wooden parts include gratings, keel, stem and sternpost and decoration.

Left: White metal detail parts. The figurehead is not a woman in riding habit but a gentleman in a tricorn that is faithful to the original Mamoli designs.

this kit is aimed at inexperienced modellers, a number of fiddly details such as stern lanterns are not included, and the fitting set is simplified to aid assembly. A piece of fine white cloth is provided to make sails, and the instructions contain templates to cut them out and instructions on how to *bend* (attach) them to the yards. The most taxing aspect of the rigging will be to glue lengths of thin thread across the shrouds to form the ratlines in this small scale. A single modern Union flag is provided for the mainmast to remind people *Bounty* is British.

As a beginner's kit this Mini Mamoli scores highly. The quality of the laser cut parts, 3D printed wooden hull, general detail and simple rigging gives the new modelmaker enough of a challenge without becoming overwhelming. In the hands of a seasoned model builder, the solid hull alone provides a great deal of potential using miniature shipbuilding techniques for a jewel-like miniature. The deck can be planked with wood shavings and trenalled, rigging from wire, bottom coppered with painted paper, and rigging blocks simulated with paper cut outs and blobs of glue. Flags and pennants from painted paper and miniature crewman converted from model railway figures, or the hull can be cut down and set into a simulated sea. The scope provided by this model is tremendous for a very modest price.

MAMOLI **HMS *BOUNTY*** 1:100 Scale

This is a new kit designed by John Gardner. This smaller scale model is double planked in lime and walnut. Again, the bulkheads and false keel are laser cut in quality birch ply and the laser cut parts primarily consist of the stern decoration and quarter galleries. The main deck is completely planked with wood strip. This kit has a few more white metal castings in which the guns and carriages are separate parts.

For rigging, both deadeyes and blocks are included, along with dowels to make the masts and yards. The mast tops are laser cut solid, but other mast parts like mast caps and trestle trees are also laser cut. An etched brass sheet provides deadeye strops, and the chains can be made from brass wire with chain plates from painted paper or brass strip as desired. The kit features several sheets of drawings and construction diagrams, and the rigging plans showing the ship in harbour rig are fairly complete for a model of this small scale. Overall, the kit is suitable for someone who wants a smaller model and considering its size the kit will yield a nicely detailed model, but gives scope for the more ambitious modeller armed with a few references. The details in this Mamoli kit are designed to be more historically accurate.

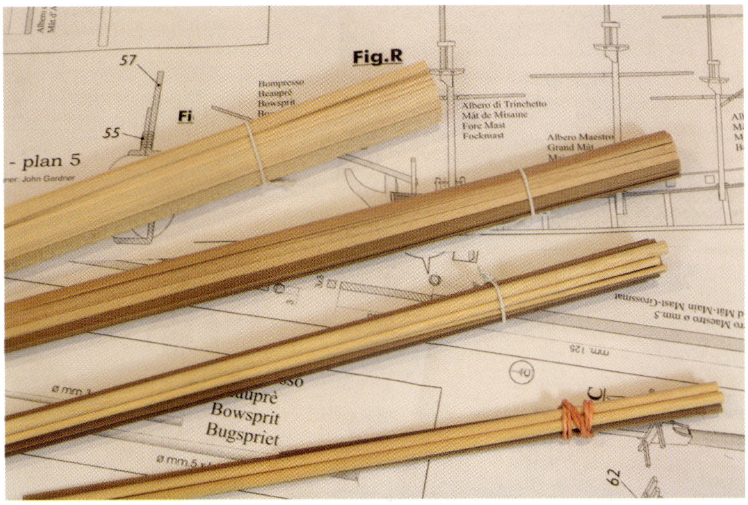

Above: Clean and clear strip wood for planking the deck and hull. The instructions are clear and concise illustrated with elevation and perspective drawings.

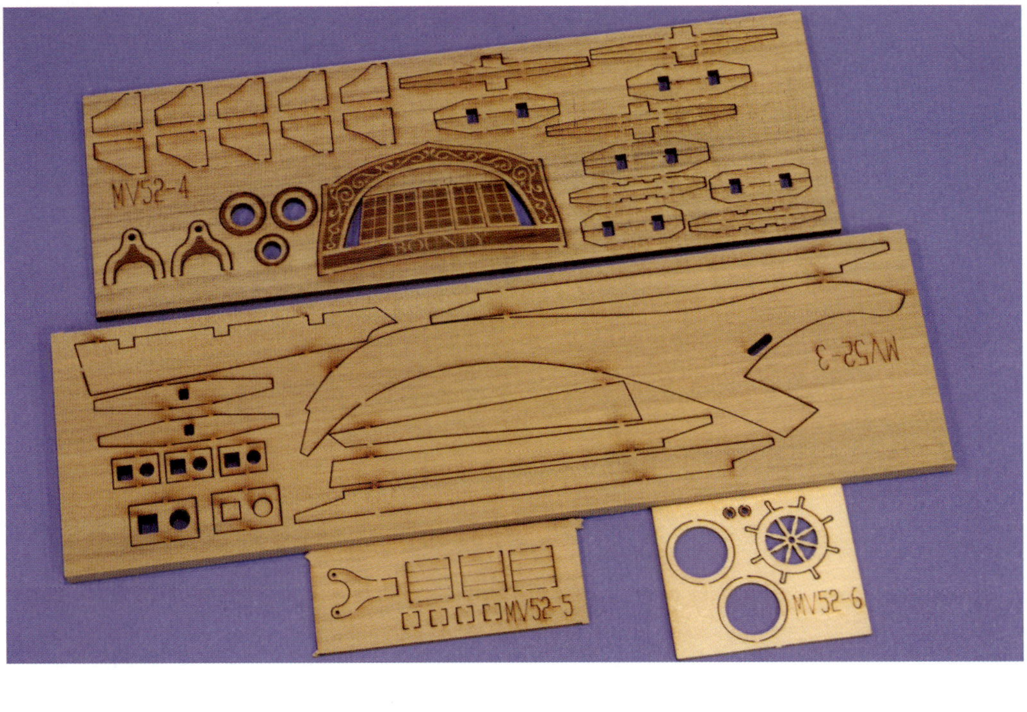

Right: Hull and masting details in walnut and ply.

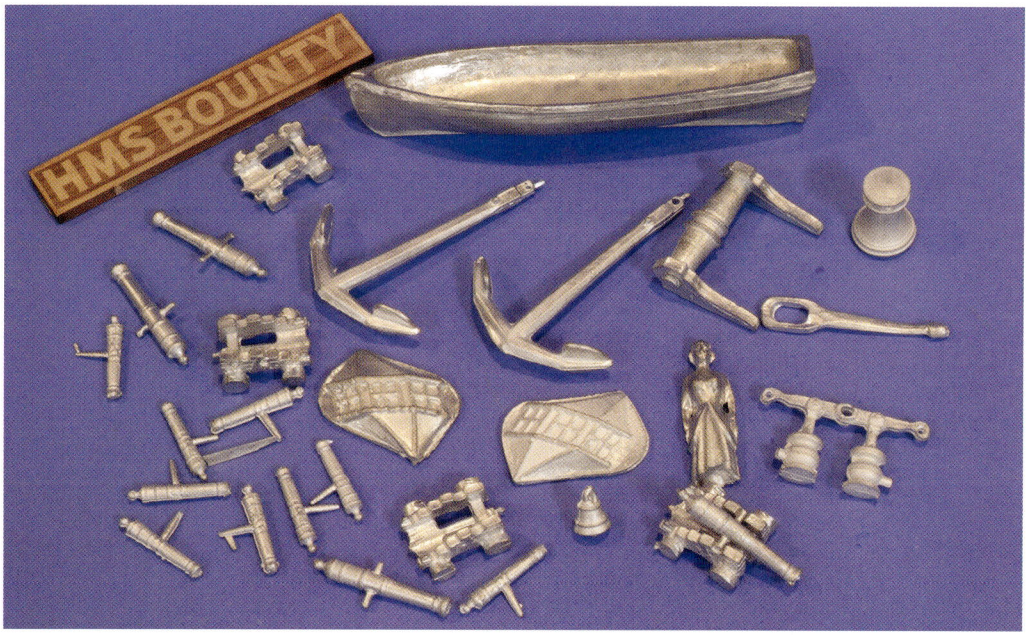

Left: Fittings are cast white metal and this time the figurehead is a woman in riding habit.

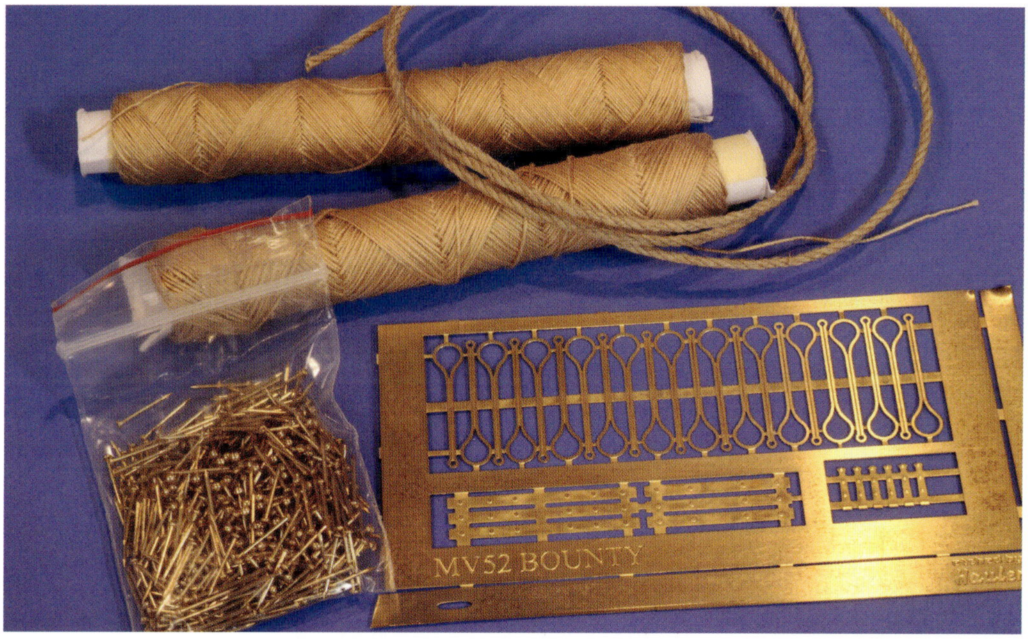

Left: Miniature brass brads, rigging cord in two sizes, anchor cable, and etched brass deadeye strops.

Left: Turned deadeyes, generic wooden blocks, belaying pins, and brass eyelets for eyebolts and jackstays. The flags are printed cloth with the correct Union, but the White Ensign should not be used.

MAMOLI HMS *BOUNTY* 1:64 Scale

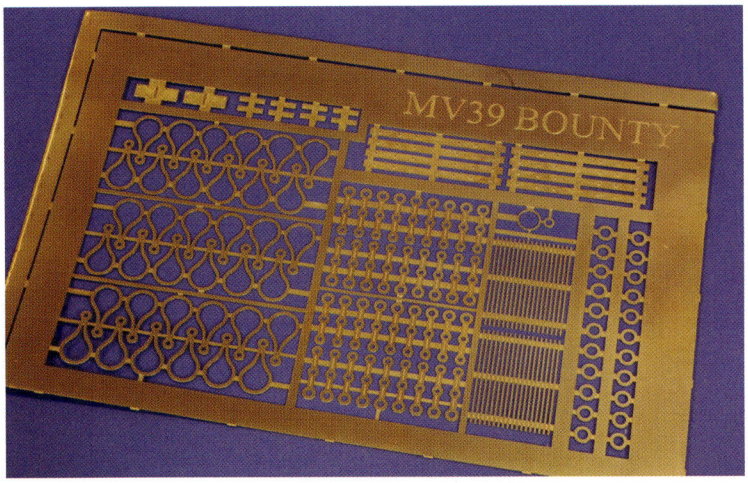

Above: A new brass etched sheet provides rigging hardware.

This large scale kit was originally designed by Luigi Volonte whose upgrade by Daniel Dusek retains the romantic features so characteristic of the brand. If you are nostalgic about the model kits of your youth you will appreciate Volonte's designs. The new kit includes all of his original plans that have a real vintage charm, given they were artistically drawn by hand and characteristic of the time; they include hints and tips about improvements one can make to the basic kit and other scratch building advice. The drawings are extremely well detailed, printed, and a real pleasure to study.

The kit is double planked but all of the parts are now cut by laser. High quality

Right: Gratings and detail cut from walnut.

Left: Rigging components, white metal launch, and flags.

Below, left: White metal detail parts. Volonte's original designs provide quite a different feel and charm to the model.

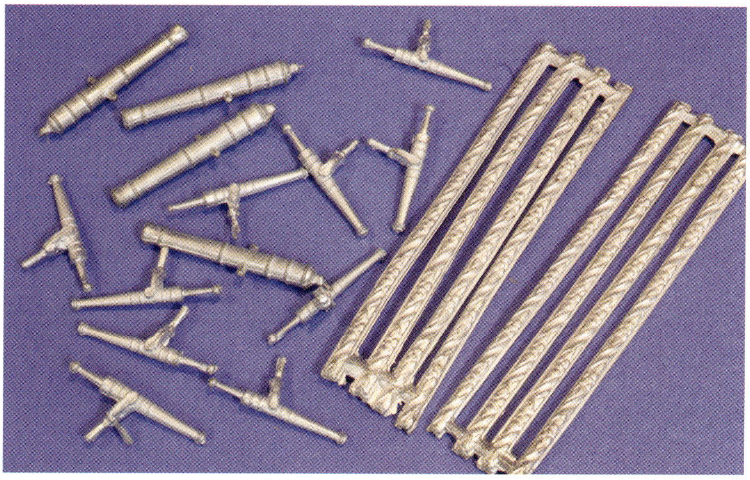

Above: White metal swivel and 4-pounder guns. Decorative strips give the model a romantic high seas appeal.

strip wood is included for the hull and deck planking, and the usual assortment of walnut and birch dowels to fashion the masts. The mast tops are solid. The castings are where you can see original romantic Volonte designs. The bow cheeks are decorated with relief carvings, and the pumps have a flourishing design. The figurehead is not a woman in riding habit, but a fellow in a tricorn who has a nautical look to him. It must be remembered that the kit was designed at a time when research materials were scarce and continental designers were trying to capture the romance of a square rigger. At this time designers had to fill in the blanks with their own ideas and continental designers added a little flair on top. With better research available, the modeller can adjust these designs to be more authentic – or it can be built the way it was designed into a delightful model. It is nice to have that choice and to see an old friend again.

MODEL SHIPWAYS HMS *BOUNTY* LAUNCH 1:16 Scale

When Bligh left Batavia he noted in his journal that he felt regrets that he could not take the launch back with him. Model Expo offers the next best thing: a remarkably accurate large scale kit designed by Bob Crane of the 23ft British naval launch. The model is constructed in much the same way as the real boat over a mould constructed from laser cut formers fixed to a stout base. The boat frames are wooden strips shaped by bending them around the formers. The keel, stem and sternposts are assembled from several parts that have all of the proper scarf joints laser cut into them. The hull is planked

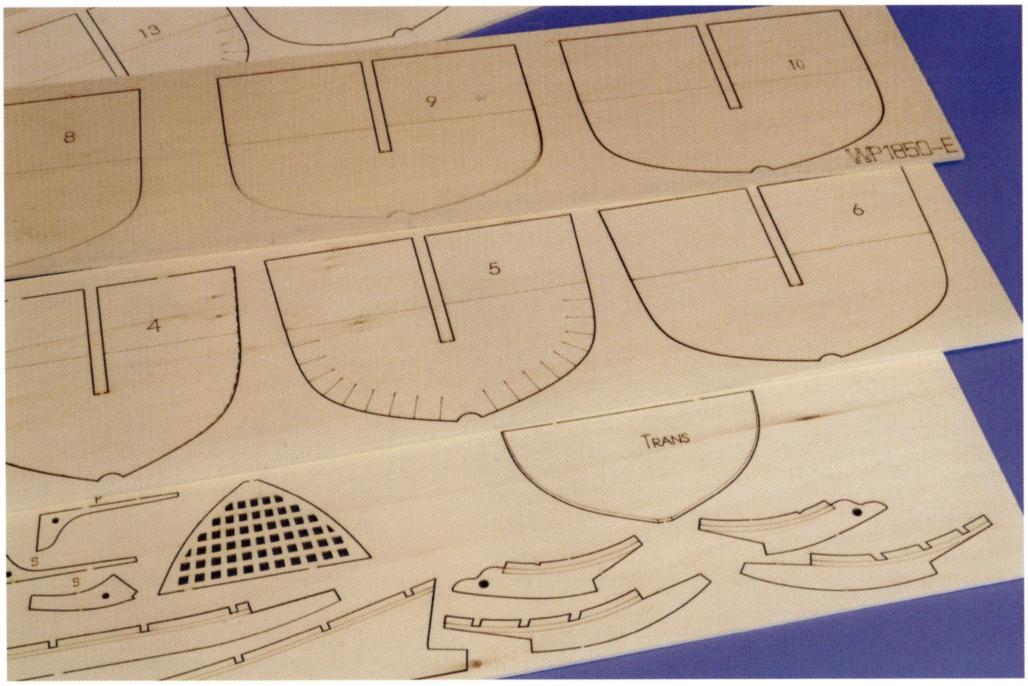

Right: Hull formers are marked with the location of each plank as they sweep up the hull ensuring accurate planking.

with ready *spiled* (shaped with tapers and curves cut into them) laser cut planks to fit the double curvature of the frames, eliminating a lot of tedious fitting. The sheets that carry these parts are even carefully marked with instructions as to what side or edge of each plank should, for example, be fixed to the keel or which edge of another plank, greatly reducing error. Once the hull is built, it is cut away from the mould and the thwarts, gunwale, and interior details are prototypically built from wood and brass strip. The kit is made from high quality basswood that is easily shaped and bent with the application of a little heat.

A real highlight is the extensive set of 1:1 scale plans of each wooden part and a well written, step by step illustrated instruction booklet. Detail fittings such as oars are laser cut blanks requiring a little shaping with sandpaper; turned casks and barrels, wooden belaying pins and a tool chest are

Above: Turned wooden barrels and scale belaying pins, white metal grapnel, and some generic rigging blocks are included. Sailcloth is a fine weave off-white cotton.

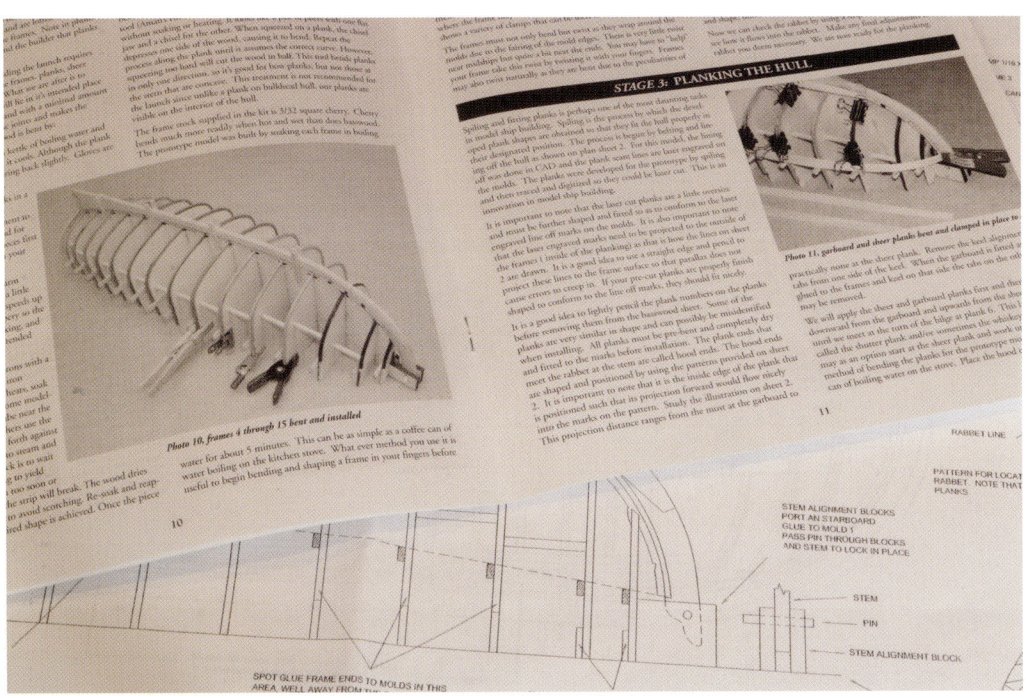

Right: Detailed written instructions supplemented with full scale plans leave no question for the builder.

all included for the interior of the boat. Fine weave cloth is provided for sails, and the instruction booklet demonstrates how to make them from laying out on the cloth, cutting them, sewing, and bending them to the yards. From reading Bligh's journal there is much else you can make from scratch and load into the boat – even the makeshift Union flag they raised as they sailed into Cupang.

Model Expo's launch kit is the most

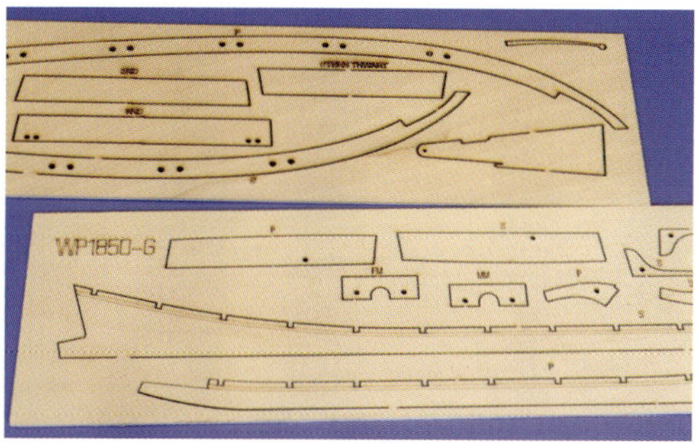

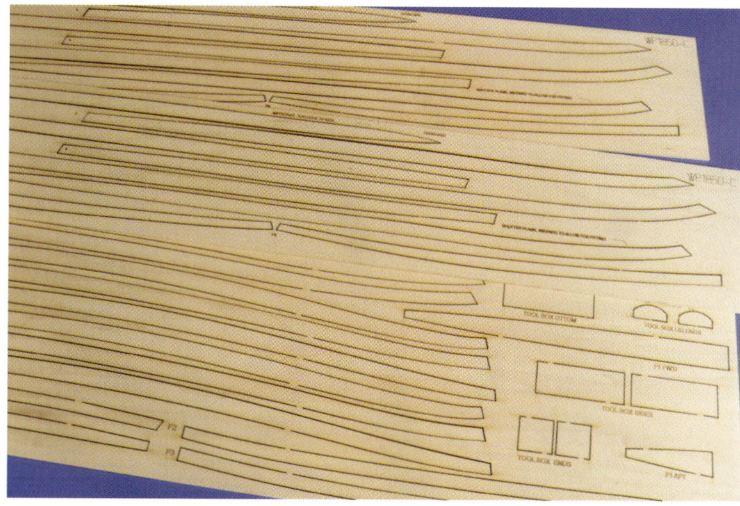

accurate model of the 23ft launch and gives the modeller a real education as to how ship's boats were made. Highly recommended.

Left: Boat keel, rails and other components are constructed with the proper scarf joints.

Above: The hull planking is pre-shaped capturing the double curvature of the hull making planking a pleasure.

HAROLD HAHN **HMS *BOUNTY*** PLANS FOR A FRAMED MODEL

1/48 Scale

A fully framed model can be scratch built using the famous 'Hahn method' illustrated on his seven-sheet plan set. The essence of the Hahn method is to glue together two thin layers of wood to form the rough shape of a frame. A pattern for the frame is glued on top of this sandwich and the frame sawn out. What makes the method unique is that the frames are glued upside down into the building jig, made possible by extending the tops of each frame so the whole hull sits absolutely level to the building board. This method makes it easy to locate waterlines, ports and other hull features because the model now has a constant datum. The method is described in his 1988 book *Ships of the American Revolution and Their Models.* The frames are stylised and do not represent actual framing practices, as is the keel, stem and stern in that not every timber (*eg*, inner and outer stem, gripe, etc) is a separate part, but rather drawn as a complete unit accurate in outline.

The plans are beautifully drawn and were based on a reconstruction of *Bounty*'s draughts. The drawings provide the general arrangements, disposition of each frame with their extensions fully drawn out, and patterns for the frame blanks. A full size building jig pattern is provided as is a rigging plan. It is not an inconsiderable task to source the wood for *Bounty* and The Lumberyard in Ohio, USA offers a timbering

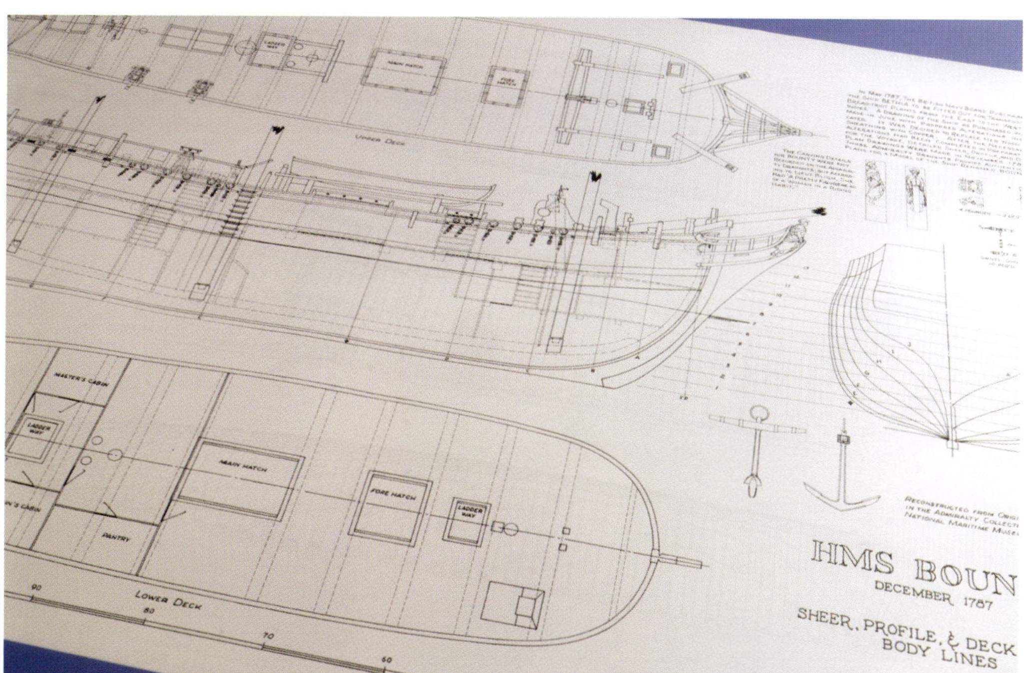

Left: General arrangement drawings.

set for these drawings. A timbering set contains all of the wood required ready milled to the sizes required. In these sets you will find billets of wood for the carvings, strips for planking cut to the correct nominal size, wooden laths to make up the frame blanks, and a laser cut building jig. Perhaps a few parts like the keel and stem parts are pre-cut for you. The value of a timbering set is that it ensures that the wood is of a consistent colour and is free of knots and figures, and saves a lot of work cutting down then milling it all yourself. Despite this convenience, *all* of fittings will need to be scratch built and all decorations carved by hand.

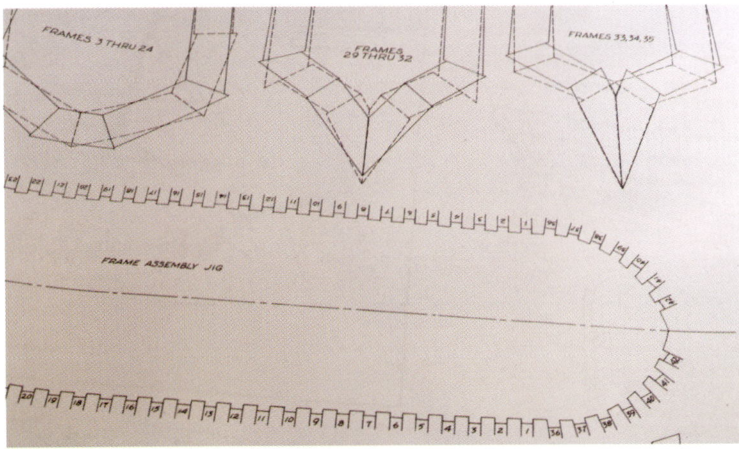

Above: Building jig and frame blank patterns.

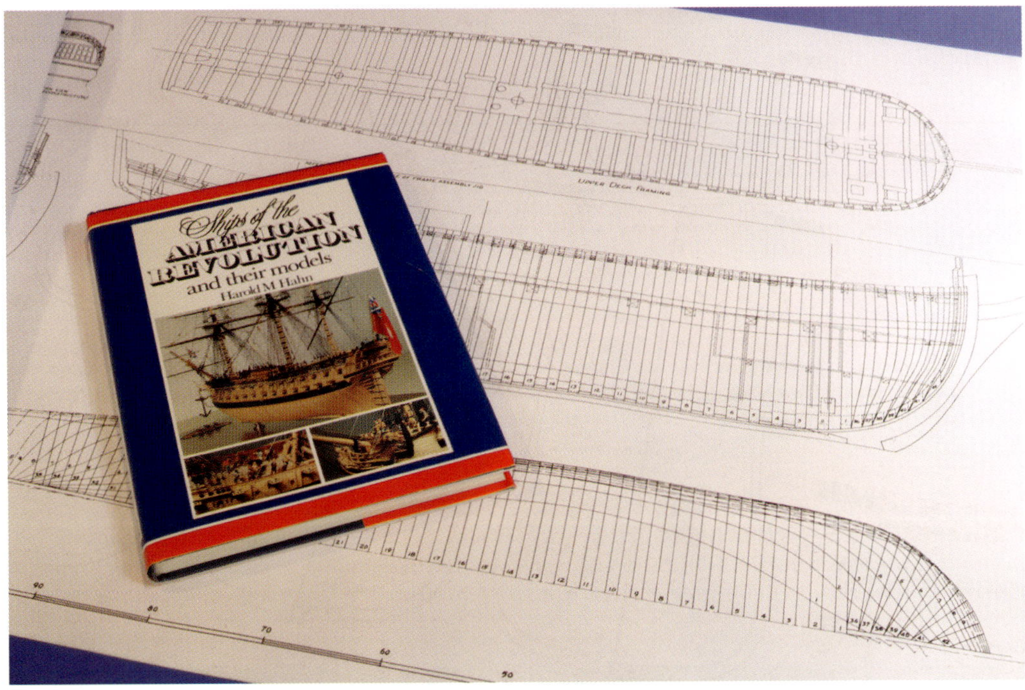

Right: Disposition of hull frames.

■ TABLE 2. SUMMARY OF OTHER READILY AVAILABLE *BOUNTY* KITS (IN ALPHABETICAL ORDER)

Manufacturer	Scale	Construction	Notes
Amati 'First Step'	1:135	Solid Hull	Designed for beginners it is simplified and toy like. Lacks the finesse of the Mamoli kit.
Artesania Latina	1:48	Single POB	Cutaway model. Gratten tops included.
Artesania Latina 'Jolly Boat'	1:25	Double POB	*Bounty* did not carry a jolly boat.
Billings	1:50	Single POB	Generally well detailed. Plastic fittings.
Constructo	1:110	Solid Hull	Designed as a beginners' kit, the model is highly simplified with crude detailing.
Constructo	1:50	Single POB	Highly simplified. No gratten tops. Designed to be more a decorative rather than authentic model.
Corel	1:130	Solid hull	A simple and toy-like beginners' kit.
Great Union	Unknown	Solid hull	The completed model resembles a seaside souvenir shop item.
Mamoli 'Bounty Launch'	1:16	Plank on Frame	This is the Model Shipway's kit. Highly recommended.
Mantua Silhouette	1:120	Single POB	Designed for beginners, the hull is plank on bulkhead that is not fully planked to emulate an 'Admiralty' model.
Minicraft	1:125	Plastic	Gold plated!
Occre 'Bounty Boat'	1:24	Double POB	Model of *Bounty*'s 23ft launch.
Pyro/Lifelike/Lindbergh	1:180	Plastic	Highly simplified plastic kit with crude details. Great for children.
Revell	1:110	Plastic	Heavily moulded details, thick plastic ratlines.
Sergal	1:60	Double POB	Sold under the 'Mantua' label as well with minor differences. The kit is very crude with out of scale fittings. Obsolete.

POB = plank on bulkhead design

Modelmakers' Showcase

AMATI **HMS *BOUNTY*** 1:60 scale

By DAVID LENDRUM

David Lendrum of Australia substantially reworked the Amati kit using John McKay's drawings. He writes that with this reference he used his scratch-building skills to build the ship's boats much more accurately, and refine a plethora of smaller but important details giving the kit a true scale detailed appearance. The model is finished akin to the movie replica ships.

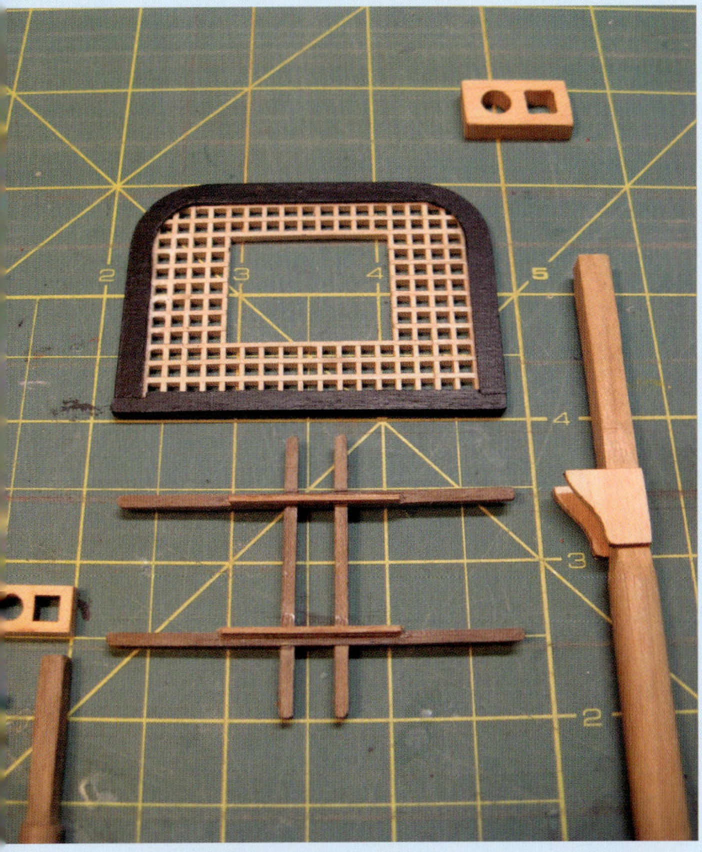

ARTESANIA LATINA **HMS *BOUNTY*** 1:48 scale

By TIM MOORE

Tim Moore of Ontario, Canada extensively detailed this popular cutaway kit to a very high standard. The ship is finished using a 'paint with wood' approach to highlight the fine hardwoods used in the model's construction.

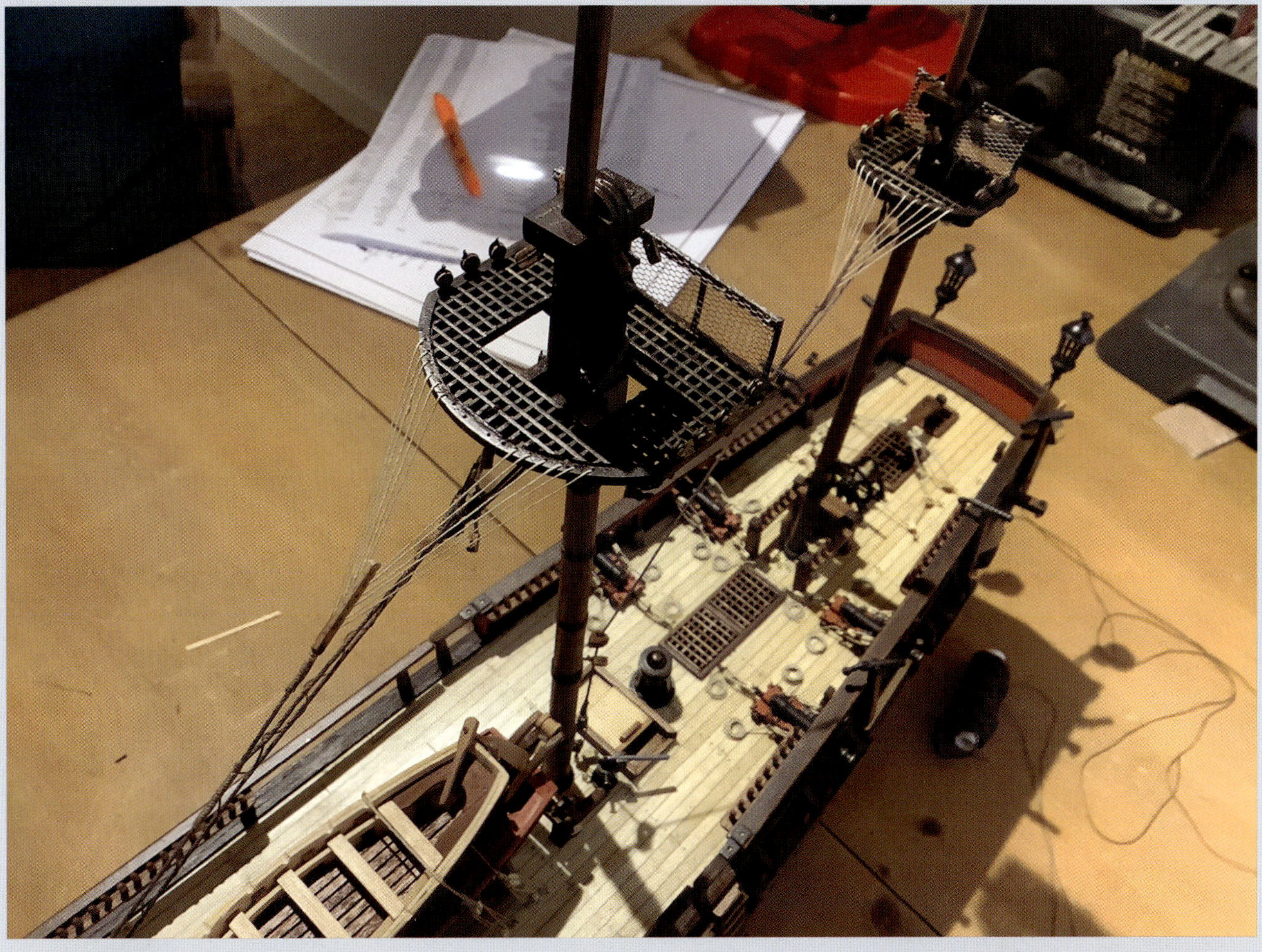

CALDERCRAFT **HMAV *BOUNTY*** 1:64 scale By KERRY JANG

This model was built out of the box as a study of the possible authentic colours of *Bounty* based on Navy Board orders, regulations, and logs from the ship and Bligh. The paint shades chosen were those probably carried in the ship's paint locker so the scheme would rely heavily on yellow and red ochre, and black.

The run of the anchor cables are incorrect and the kit instructions should not be followed.

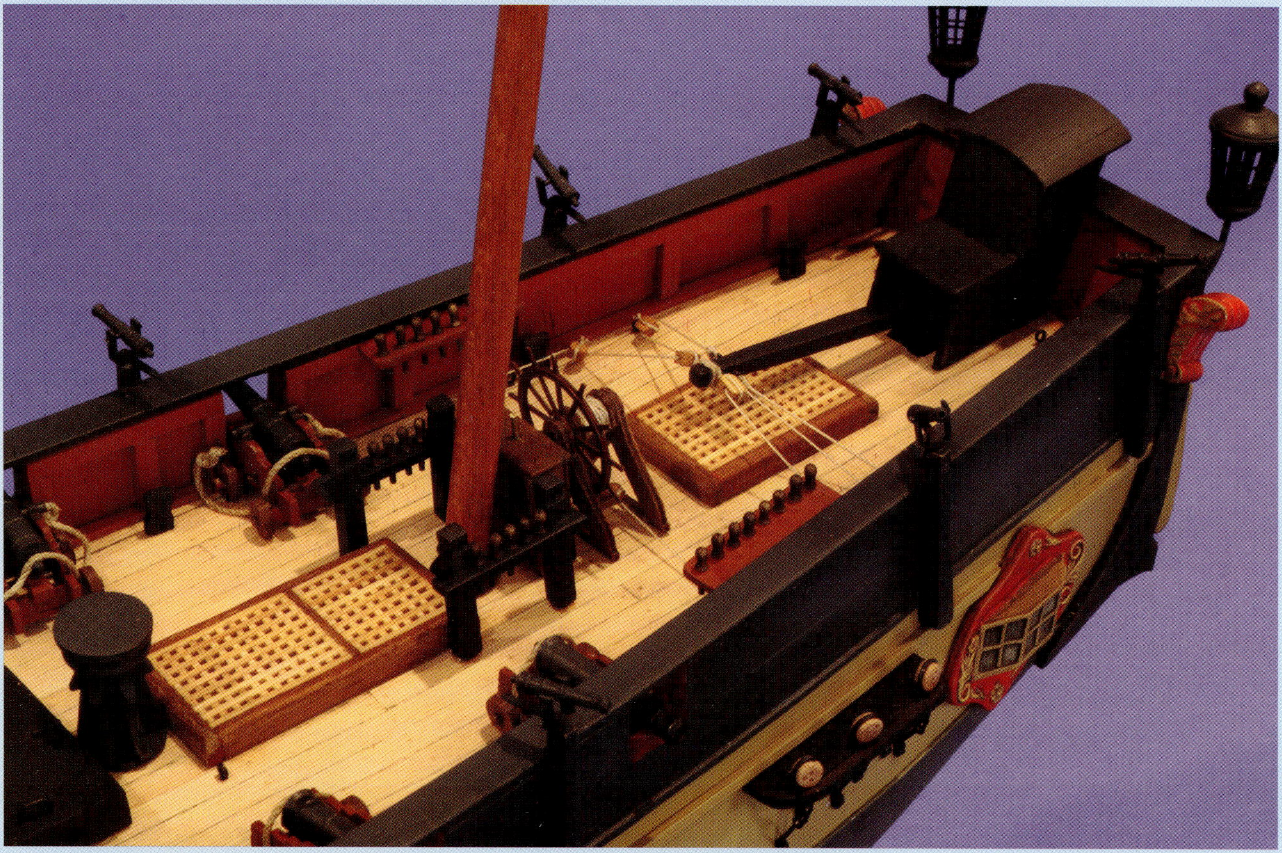

DEAGOSTINI **HMS *BOUNTY*** 1:48 scale

By KERRY JANG

The DeAgositini 'Admiralty Style' model was modified to omit more of the hull planking in order to allow a clearer view of the interior framing. The main deck was planked to a greater degree for personal aesthetic preferences as was the use of paint on the hull sides and for many of the deck fittings, in keeping with the style of Navy Board type models.

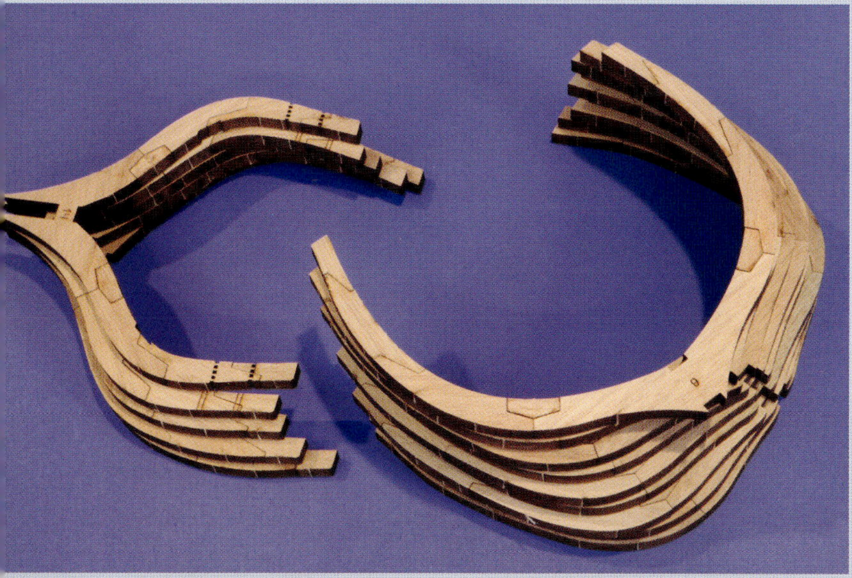

Appearance

Two scholarly papers – 'The Application and Scheme of Paintworks in British Men-of-War in the Late Eighteenth and Early Nineteenth Centuries' by Peter Goodwin, and 'Pitch, Paint, Varnish and the Changing Colour Schemes of Royal Navy Warships, 1775–1815: A summary of existing knowledge' by Brian Vale published in *The Mariner's Mirror*, help us determine 'what Bounty did not look like' and reconstruct how she may have appeared.

The first major change to *Bethia* was to copper her hull; this and the conversion of the ship was carried out by the Navy and her finish most likely followed the painting instructions of 1780. These specified the ship's sides are to be painted *yellow* (yellow ochre) above the main wale, with the wale and below to the top of the copper in black. This instruction is qualified by 'as desired by the commander' allowing for some variation. As for the shade of yellow, contemporary accounts of British ships noted that the yellow provided by the Navy Board '... is so very dark a colour that it makes the ship look very dirty and dismal.' There are several references to *Bounty* being painted in logs and journals, and no reference to her ever being *payed* – having her hull timbers finished in a composition of pine vanish, turpentine and rosin that gives the ship's side a shiny light mahogany appearance.

The instructions also specify the inside weather-deck bulwarks to be painted *red* (red ochre) and interior spaces of the ship *whitewashed* in a mixture of fish glue and zinc oxide thought to absorb odours that cause disease among the crew. The holds could be whitewashed or painted grey, typical of English practice as noted by the French shipwright Blaise Ollivier.

The ground colour of *Bounty*'s upper outer bulwarks is unknown. The Navy Board instructions prescribe black, but in contemporary models and paintings of other ships they are frequently dark blue. Navy Board instructions permit colours other than those specified to be used at the discretion of the captain (and his pocket) so *Bounty's* hull could have had them painted blue, but Bligh does not make any mention of this in his logs or when her paintwork was repaired and refreshed during the voyage. *Bounty* was rated a schooner and if we take this as a lead, contemporary illustrations of small vessels show the ground as black, with rated vessels displaying blue with red accents, or black with other coloured accents.

The shape and form of the *Bounty*'s stern, windows, cove and transom are shown in the draughts, but it is not clear what her carved decoration, if any, was like. There exists one illustration by Dodd from 1790 that shows a decorative cove above the windows with carvings. The colours of the stern's cove are also unknown. Bligh has been described as being quite thrifty with personal funds so he could have used what was available in the ship's stores such as red ochre, yellow ochre, and black. With this palette the stern could have been black with any decoration picked out in yellow ochre, or red with yellow ochre details. The same would apply to the side window quarter badges.

Bligh described the figurehead as a 'woman in riding habit' but the appearance of the actual carving is unknown. What is known is that Bligh recorded that he had the figurehead painted in *bright colours* during their stay at Otaheite much to the delight and approval of the Otaheitians. This suggests that the figurehead was originally painted white that accords well with David Pulvertaft's article 'The Colour Schemes of British Warship Figureheads 1727–1900' published in *The Mariner's Mirror* that found most ships' figureheads were white and later painted by the crews and captain. Any colours suggested for the figurehead's repainting would be conjecture, but study of illustrations of English riding habits at this time suggests they were predominantly red with black trim

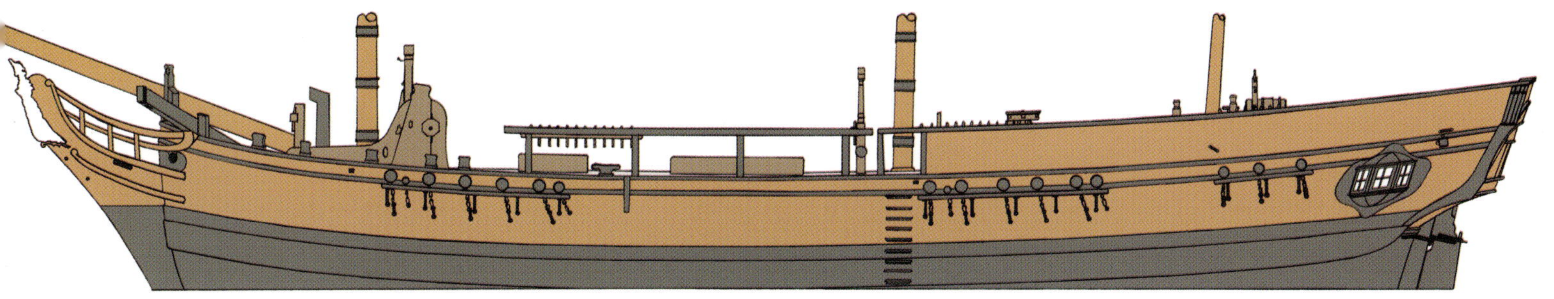

Bethia (1787).
The profile is based on contemporary illustrations of colliers and a model of *Earl of Pembroke*, a collier that was converted to James Cook's *Endeavour*. *Bethia* is in a payed finish with black lower hull (note that the black is rendered as dark grey in the illustrations so as not to obscure details). The fittings are shown in a varnished wood finish. Colour accents could have been used for the stern and quarter badges but any decoration would have been generally modest for a working vessel.

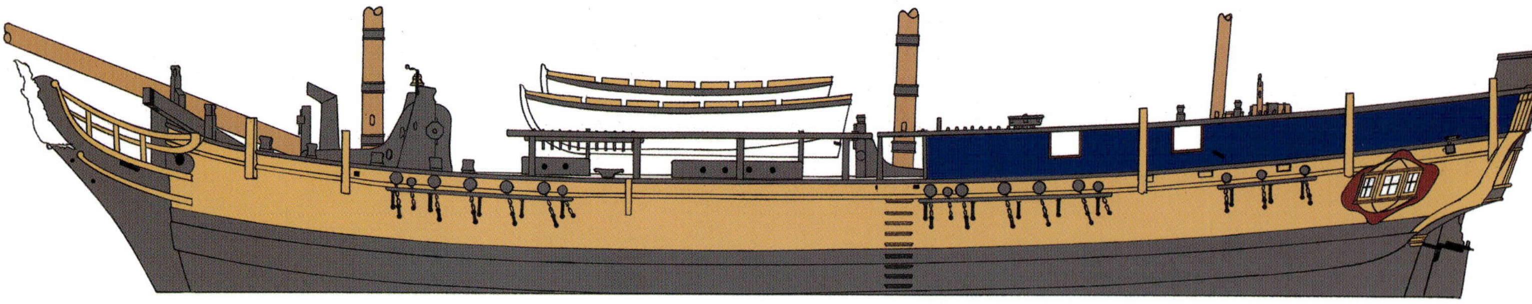

HM Armed Vessel *Bounty* (1787). This profile used colours based on Navy Board instructions where black and yellow ochre predominate. The ground for the upper outer bulwarks are shown in blue after contemporary models of other ships, but could have just as easily been black. The swivel gun posts are shown in ochre, but could have been black as well. Inner weather bulwarks are red ochre. Deck fittings and structures are shown in black (could have been left in a varnished finish) with lower masts, wheel and binnacle in varnished wood.

topped with a tricorn hat. There is a surprising amount of information on riding habits because they were controversial at the time because they were considered 'unfeminine' (by the men in the male dominated society) and being tailored like men's clothing. Certainly other colours were used such as green or light blue but museum samples suggest these colours were more the fashion with the ladies of the continent.

Bounty's hatches were built or modified to have ventilation holes on the sides of the coamings. These hatches were probably painted black or red ochre to protect their surfaces. The capping rails and stanchions would have been black, which is typical of the time. The swivel gun posts could be black or yellow ochre; there are no recorded details. Gun carriages were furnished to ships in red ochre and only with written permission could they be repainted in different colours (no such request by Bligh can be found).

The masts and bowsprit were probably varnished. There is no evidence whether they were repainted and Navy Board instructions are silent on the colouring of the lower masts and bowsprit. However, the Navy Board instructions do specify that the tops, upper masts and yards were to be blackened, and in 1779 ordered that black varnish should replace black paint because varnish conferred greater waterproofing.

Ship's boats were provided by contractors to standard designs. The Navy Board instruction on the finish of boats is that they are to be painted white with black wales, and interiors painted yellow ochre. The use of other colours depended on the captain's purse with colours like green typically reserved for flag officers.

Bounty's name was likely painted on her stern in white, and examples of their style are shown on many 'as fitted' draughts of the 1770s and 1780s.

As with *Bounty*, there are no contemporary illustrations of *Bethia*. There is no evidence that she was anything but typical of a collier of the time, and contemporary models, such as the *Earl of Pembroke* that was purchased and converted to Cook's *Endeavour*, show a payed finish with black wales. Her underwater hull was sheathed and possibly also covered in *brown stuff* (a mixture of pitch and sulphur to discourage worm). Contemporary paintings and illustrations of merchant colliers show a similar finish and that the masts were left in varnished wood. Yards and tops were probably black but could have been left natural wood and varnished. Decoration was simple with a coloured ground with any carvings picked out in paint.

Bounty would have flown a Red Ensign, the usual distinguishing mark of British warships (and all merchantmen for that matter) not assigned specifically to a squadron commanded by an admiral of the Blue or White. The Red Ensign would have been a 1742 pattern with a 5 to 9 breadth versus length proportions that carried a 1606 Pattern Union in the canton. It would have been flown from an ensign staff or halyard at the stern, and a red commissioning pendant from the main masthead. Other than signal flags, she would wear no other flags except for a Union Jack from the bow whilst at anchor. *Bounty* was flush-decked and 'unrated', and it is unlikely that the practice of sometimes flying a Union from the fore masthead would have been applied to so small a ship.

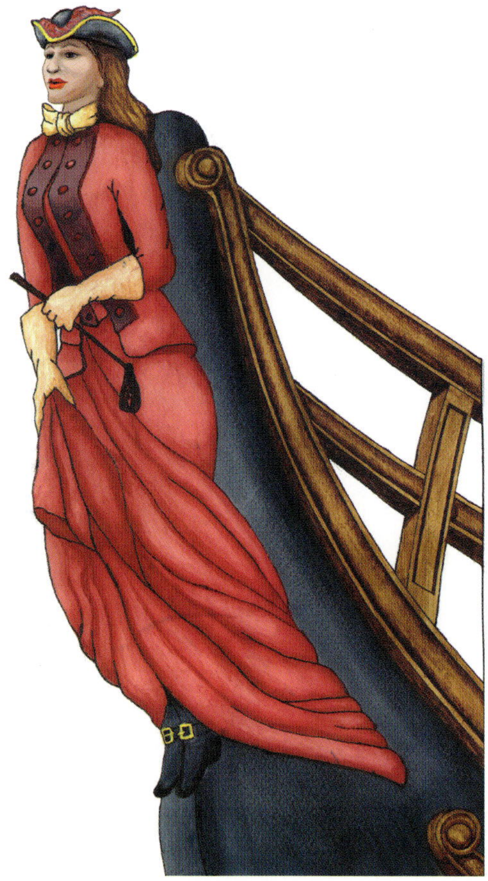

Figurehead.
All that is known about *Bounty*'s figurehead is that it was a woman in riding habit and that she was painted in 'bright colours' at Otahteite. The illustration is based on Georgian riding habits held in the Victoria & Albert Museum and contemporary illustrations of prominent women such as Lady Seymour Worsley. At *Bethia*'s building and *Bounty*'s commissioning the figurehead was probably painted white.

Bethia (1787).
Stern.

Stern Decoration.
There are no contemporary illustrations or description of the stern decoration. Three possible sterns are presented. Top is for *Bethia* following the idea colliers had little or no decoration. Below left is *Bounty* with decoration taken from the replica ships. The ground colour is shown as blue but just as likely is black or red ochre with trim in ochre typically found on contemporary models. Below right is *Bounty* with no stern decoration, simply *Bethia*'s stern picked out in different colours.

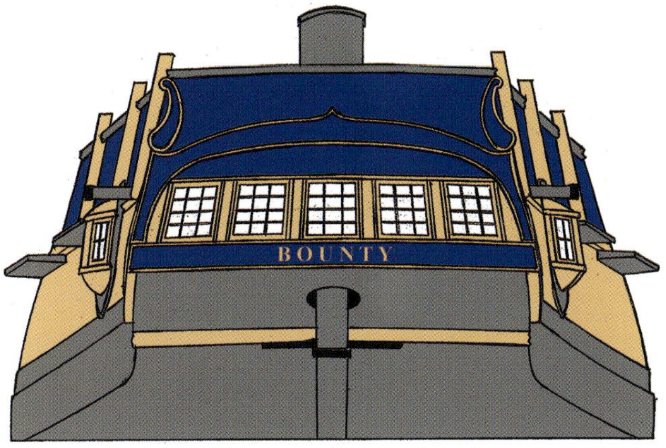

HM Armed Vessel *Bounty* (1787).
Stern (with decoration).

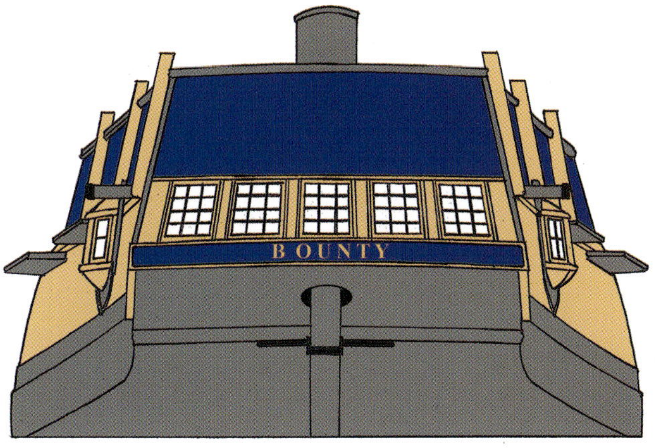

HM Armed Vessel *Bounty* (1787).
Stern (without decoration).

Stern Decoration Detail.
There are no contemporary descriptions or illustrations of any of *Bounty*'s stern decoration. Left shows an enlargement of the decoration painted by Robert Dodd. He was reputed to strive for accuracy in his illustrations but there is no evidence as to whether his illustration was based on any sources or simply conjectural. Right is the decoration based on the replica ship built in 1962.

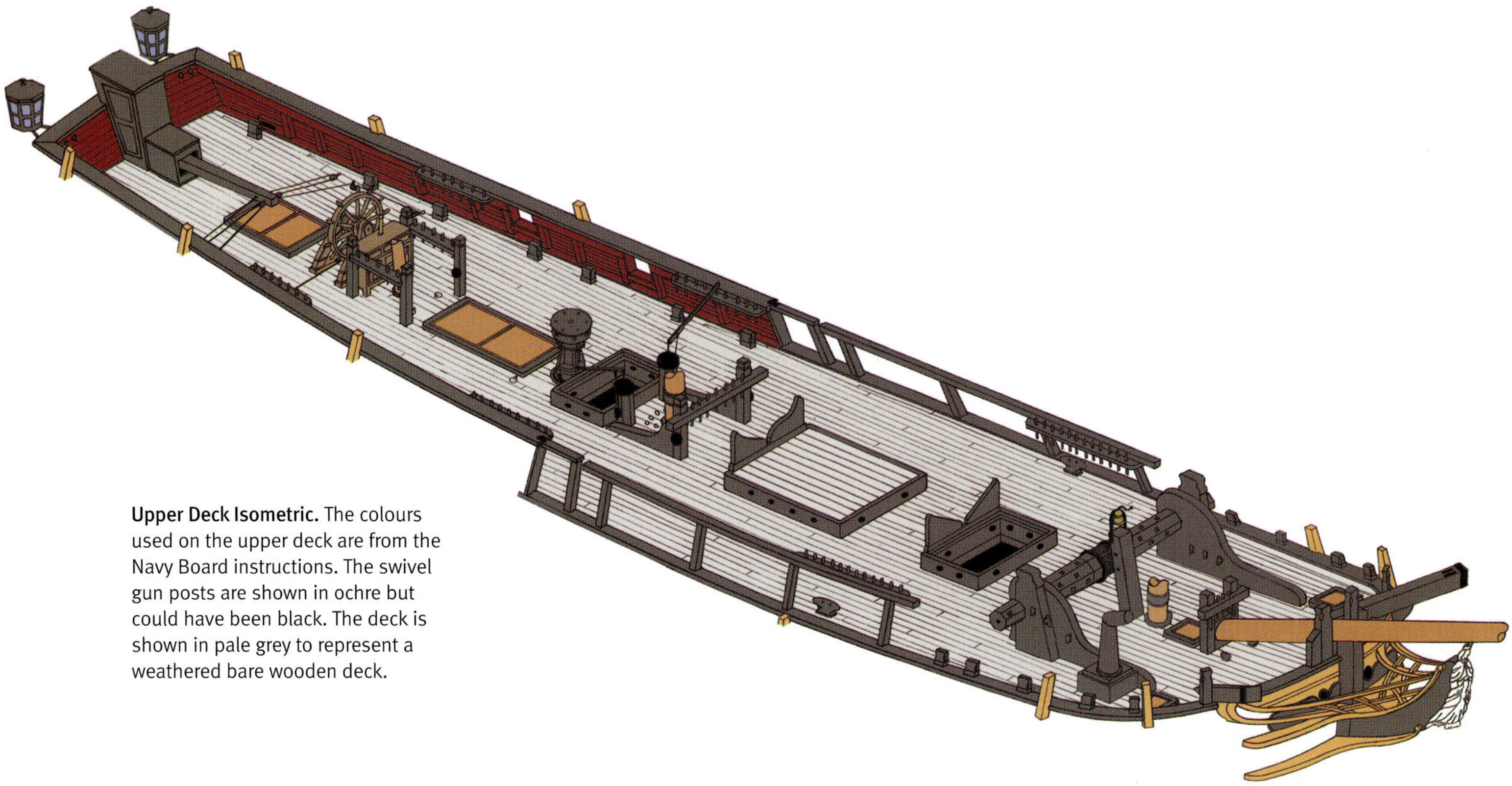

Upper Deck Isometric. The colours used on the upper deck are from the Navy Board instructions. The swivel gun posts are shown in ochre but could have been black. The deck is shown in pale grey to represent a weathered bare wooden deck.

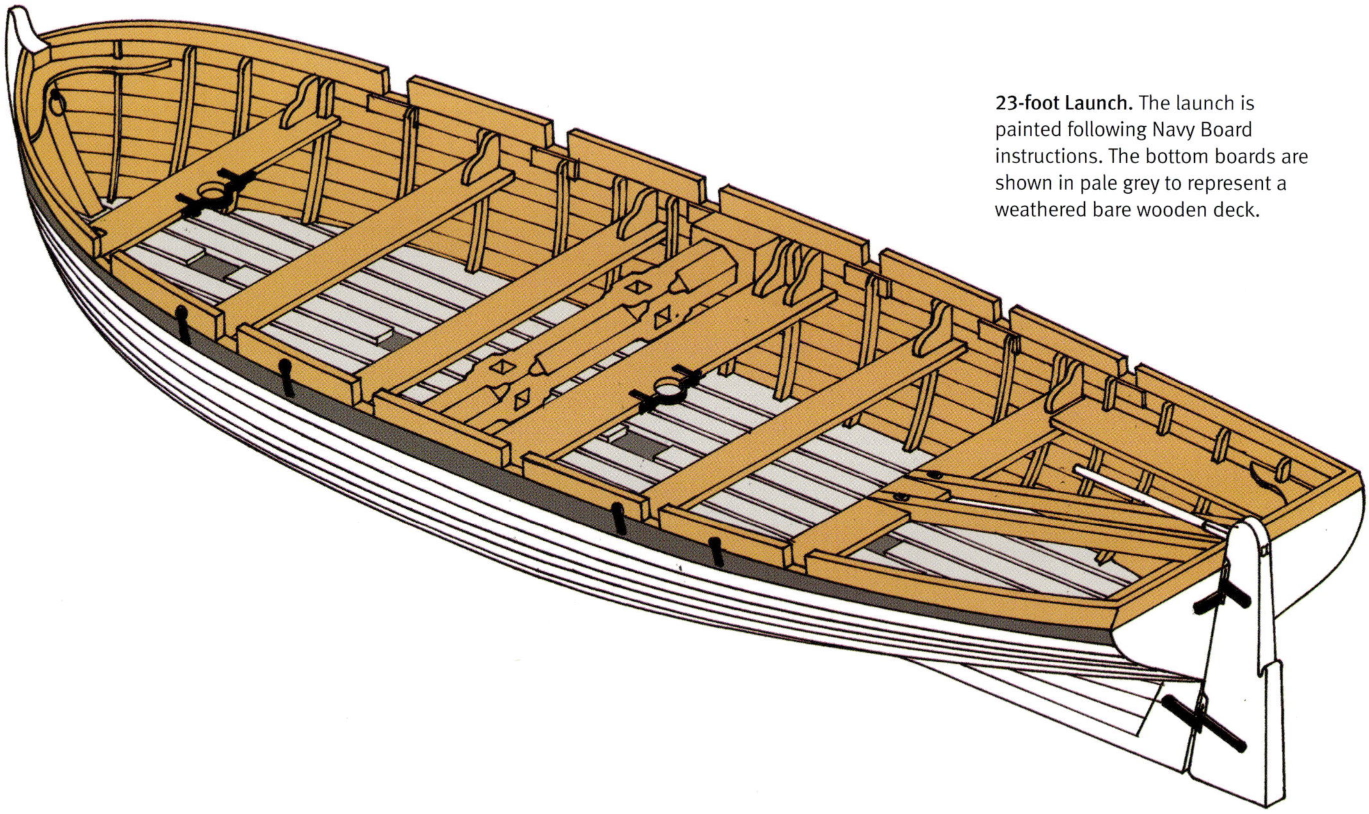

23-foot Launch. The launch is painted following Navy Board instructions. The bottom boards are shown in pale grey to represent a weathered bare wooden deck.

Masts, yards and gratten top. When *Bethia* was purchased, her masts and yards were surveyed and general dimensions recorded. *Bethia*'s masts and yards were retained for *Bounty* and were modified by the Admiralty, and further by Bligh during his tenure. The modelmaker should be aware that the dimensions of the masts and yards specified in model kits differ enormously between manufacturers. Table 3 provides in some detail the known dimensions for each mast and yard and that can be scaled to fit your model.

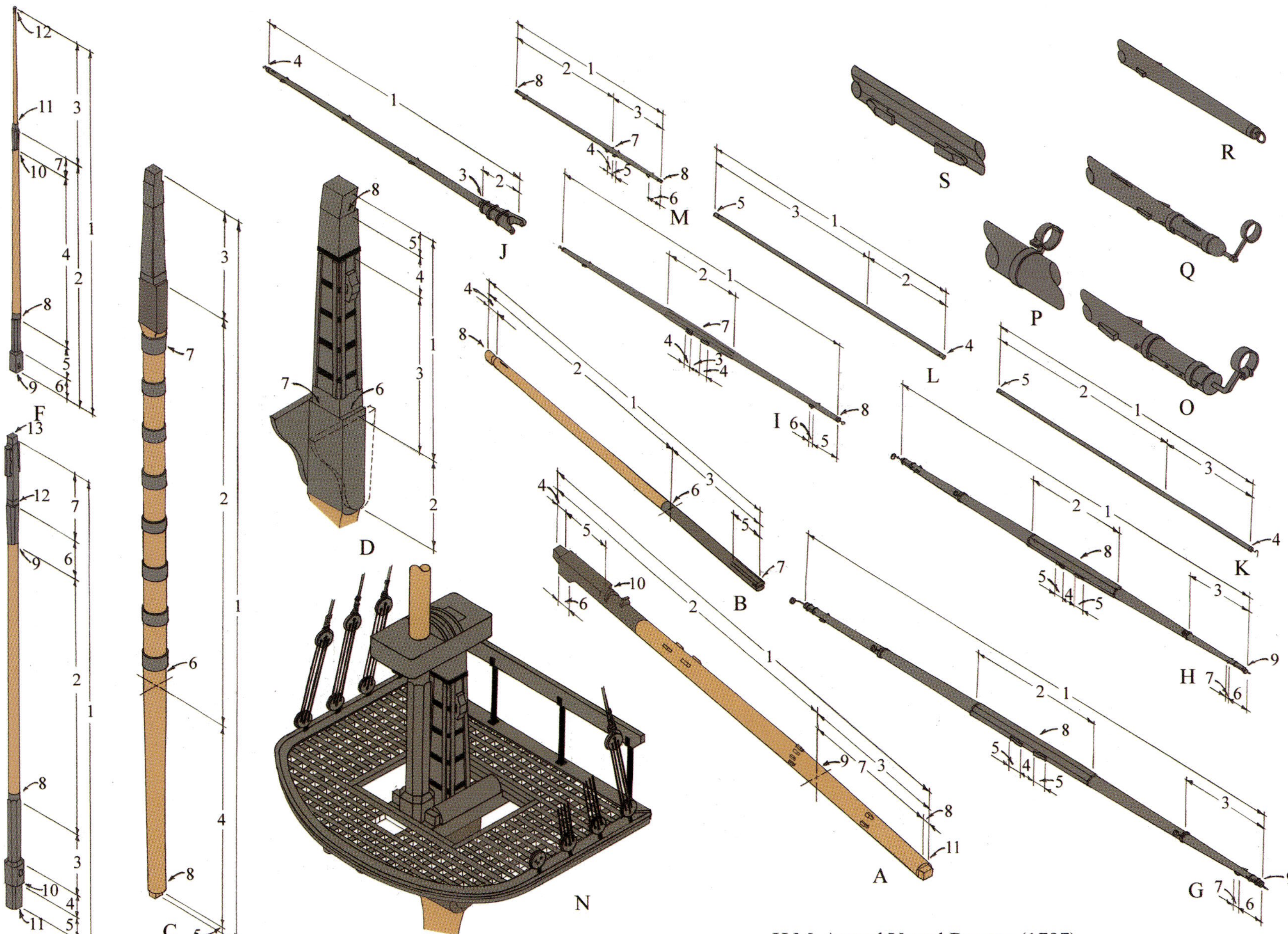

H.M. Armed Vessel Bounty (1787)
Masts and Yards

MASTS AND YARDS:
A Bowsprit
B Jib boom
C Lower mast
D Lower mast head
E Topmast
F Topmast/Topgallant mast
G Lower yard
H Topyard
I Topgallant yard/Topyard/
Spritsail yard/Crossjack yard
J Mizzen yard (Gaff)
K Lower studdung sail boom
L Topsail and Topgallant
Studding sail booms
M Studding sail yards
DETAILS:
N Main top
O Lower yard arm
P Studding sail boom iron
Q Togallant yard arm
R Yard arm
S Sling cleats

TABLE 3. MASTS AND SPARS

Item	As purchased with *Bethia*	As ordered by the Admiralty	As requested by Bligh
BOWSPRIT (A)			
1 Length	35ft 4in (10.464m)	Unchanged	Unchanged
2 Length above housing	25ft 2in (7.464m)	Unchanged	Unchanged
3 Housing	10ft 2in (3.0m)	Unchanged	Unchanged
4 Tennon	8¼in (21cm) Assumed	Unchanged	Unchanged
5 Long square	3ft 11in (1.193m) Assumed	Unchanged	Unchanged
6 Short square	1ft 3⅝in (39.7cm) Assumed	Unchanged	Unchanged
7. Given diameter to tennon	9ft 6in (2.896m) Assumed	Unchanged	Unchanged
8 Tennon	8in (20.3cm) Assumed	Unchanged	Unchanged
9 Given diameter	18¼in (46.3cm)	Unchanged	Unchanged
10 Diameter at square	12½in (31.8cm) Assumed	Unchanged	Unchanged
11 Diameter at heel	15⅝in (39.7cm) Assumed	Unchanged	Unchanged
JIB BOOM (B)			
1 Length	27ft 1in (8.255m)	Unchanged	Unchanged
2 Length above given diameter	18ft 1in (5.518m) Assumed	Unchanged	Unchanged
3. Length below given diameter	9ft 0in (2.743m) Assumed	Unchanged	Unchanged
4. End to score shoulder	13⅛in (33.3cm) Assumed	Unchanged	Unchanged
5 Eight sided	2ft 6⅝in (77.8cm) Assumed	Unchanged	Unchanged
6 Given diameter and 7 diameter at heel	8¾in (22.2 cm) Assumed	Unchanged	Unchanged
8 Diameter at outer end	5⅞in (14.9cm) Assumed	Unchanged	Unchanged
FORE LOWER MAST (C)			
1 Length	53ft 4in (16.256m)	Unchanged	51ft 0in (15.545m)
2 Given diameter to head	28ft 4in (8.635m) Assumed	Unchanged	26ft 0in (15.545) Assumed
3 Head	7ft 4in (2.225m) Assumed	Unchanged	Unchanged
4 Housing	14ft 10in (4.521m) Assumed	Unchanged	Unchanged
5 Tennon	6in (15cm) Assumed	Unchanged	Unchanged
6 Given diameter	16½in (42.91cm)	Unchanged	Unchanged
7 Diameter at hounds	14in (35.56cm) Assumed	Unchanged	Unchanged
8 Diameter at heel	14⅛in (35.88cm) Assumed	Unchanged	Unchanged
FORE LOWER MAST HEAD (D)			
1 Mast head length	7ft 4in (2.225m) Assumed	Unchanged	Unchanged
2 Hounds	3ft 5in (1.041m) Assumed	Unchanged	Unchanged
3 Hounds to iron hoops	5ft 1in (1.549m) Assumed	Unchanged	Unchanged
4 Above iron hoops	1ft 5½in (44.45cm) Assumed	Unchanged	Unchanged
5 Tennon	9½in (24.13cm)	Unchanged	Unchanged
6 Head at hounds	12⅜in (31.43cm)	Unchanged	Unchanged
7 Head at hounds	14⅛in (35.88cm)	Unchanged	Unchanged
8 Tennon (square)	10⅜in (26.35cm)	Unchanged	Unchanged
MAIN LOWER MAST (C)			
1 Length	58ft 10in (17.678m)	Unchanged	55ft 0in (16.764)
2 Given diameter to head Assumed	30ft 10in (9.144m) Assumed	Unchanged	27ft 0in (8.23m)
3 Head	8ft 2 in (2.489m) Assumed	Unchanged	Unchanged
4 Housing	15ft 6in (4.724m)	Unchanged	Unchanged
5 Tennon	6in (15cm) Assumed	Unchanged	Unchanged
6 Given diameter	16in (40.64cm)	Unchanged	Unchanged
7 Diameter at hounds	13½in (34.29cm) Assumed	Unchanged	Unchanged
8 Diameter at heel	13¾in (34.93cm) Assumed	Unchanged	Unchanged
MAIN LOWER MAST HEAD (D)			
1 Mast head length	8ft 2in (2.489m) Assumed	Unchanged	Unchanged
2 Hounds	3ft 9¾in (1.163m) Assumed	Unchanged	Unchanged
3 Hounds to iron hoops	6ft 1in (1.854m) Assumed	Unchanged	Unchanged
4 Above iron hoops	1ft 3in (38.1cm) Assumed	Unchanged	Unchanged
5 Tennon	10in (25.4cm) Assumed	Unchanged	Unchanged
6 Head at hounds	12in (30.48cm) Assumed	Unchanged	Unchanged
7 Head at hounds	13¾in (34.93cm) Assumed	Unchanged	Unchanged
8 Tennon (square)	10¼in (26.04cm) Assumed	Unchanged	Unchanged
MIZZEN LOWER MAST (C)			
1 Length	48ft 2in (14.681m)	Unchanged	Unchanged
2 Given diameter to head	27ft 3¼in (8.312m) Assumed	Unchanged	Unchanged
3 Head	5ft 4in (1.625m) Assumed	Unchanged	Unchanged
4 Housing	15ft 0¾in (4.592m) Assumed	Unchanged	Unchanged
5 Tennon	6in (15cm) Assumed	Unchanged	Unchanged

Item	As purchased with *Bethia*	As ordered by the Admiralty	As requested by Bligh
Mizzen lower mast continued			
6 Given diameter	11 1/4in (28.58) Assumed	Unchanged	Unchanged
7 Diameter at hounds	9½in (24.13cm)	Unchanged	Unchanged
8 Diameter at heel	9 5.8in (24.47cm)	Unchanged	Unchanged
MIZZEN LOWER MAST HEAD (D)			
1 Mast head length	5ft 4in (1.625m)	Unchanged	Unchanged
2 Hounds	2ft 5⅞in (75.88cm) Assumed	Unchanged	Unchanged
3 and 4 total – no Iron Hoops	4ft 9⅞in (1.213m) Assumed	Unchanged	Unchanged
5 Tennon	6⅛in (15.56cm) Assumed	Unchanged	Unchanged
6 Head at hounds	8½in (21.59cm) Assumed	Unchanged	Unchanged
7 Head at hounds	9⅝in 24.45cm) Assumed	Unchanged	Unchanged
8 Tennon (square)	7in (17.78cm)	Unchanged	Unchanged
FORE TOPMAST (E)			
1 Length	34ft 3in (10.439m)	Unchanged	Unchanged
2 Top of cap to hounds	19ft 2⅛in (5.845m)	Unchanged	Unchanged
3 Heeling to top of cap	5ft 1⅛in (1.553m) Assumed	Unchanged	Unchanged
4 Heeling	2ft 2⅞in (67.75cm) Assumed	Unchanged	Unchanged
5 Block	1ft 4⅛in (40.96cm) Assumed	Unchanged	Unchanged
6 Hounds	2ft 4¾in (73.03cm)	Unchanged	Unchanged
7 Head	4ft 0in (1.219m)	Unchanged	Unchanged
8 Given diameter	10¾in (27.38cm)	Unchanged	Unchanged
9 Diameter at hounds	8¾in (22.23cm) Assumed	Unchanged	Unchanged
10 Heeling (square)	12¾in (32.39cm) Assumed	Unchanged	Unchanged
11 Block (square}	10¾in (26.29cm) Assumed	Unchanged	Unchanged
12 Head (square)	7½in (19.75cm) Assumed	Unchanged	Unchanged
13 Tennon (square)	5¾in (14.68cm) Assumed	Unchanged	Unchanged
MAIN TOPMAST (E)			
1 Length	35ft 10in (10.922m)	Unchanged	Unchanged
2 Top of cap to hounds	18ft 8⅝in (5.705m) Assumed	Unchanged	Unchanged
3 Heeling to top of cap	5ft 11⅛in (1.807m)	Unchanged	Unchanged
4 Heeling	2ft 2⅞in (68.26cm) Assumed	Unchanged	Unchanged
5 Block	1ft 4⅛in (40.96cm) Assumed	Unchanged	Unchanged
6 Hounds	2ft 10¼in (86.94cm) Assumed	Unchanged	Unchanged
7 Head	4ft 9in (1.448m) Assumed	Unchanged	Unchanged
8 Given diameter	10¾in (27.38cm)	Unchanged	Unchanged
9 Diameter at hounds	9¼in (23.5cm) Assumed	Unchanged	Unchanged
10 Heeling (square)	12¾in (32.39cm) Assumed	Unchanged	Unchanged
11 Block (square)	10¾in (27.31cm) Assumed	Unchanged	Unchanged
12 Head (square)	7½in (19.75cm) Assumed	Unchanged	Unchanged
13 Tennon (square)	5⅞in (14.41cm)	Unchanged	Unchanged
MIZZEN TOPMAST (F)			
1 Length overall	24ft 3½in (7.404m)	Unchanged	Unchanged
2 Topmast	21ft 2½in (6.464m) Assumed		
3 Head (pole head)	3ft 1in (93.98cm)	Unchanged	Unchanged
4 Top of cap to hounds	14ft 0¼in (4.274m) Assumed	Unchanged	Unchanged
5 Heeling to top of cap	4ft 0in (1.219m) Assumed	Unchanged	Unchanged
6 Heeling	1ft 4in 40.64cm) Assumed	Unchanged	Unchanged
7 Hounds	1ft 10¼in (56.51cm) Assumed	Unchanged	Unchanged
8 Given diameter	6⅞in (17.46cm)	Unchanged	Unchanged
9 Heeling (square)	7½in (19.75cm) Assumed	Unchanged	Unchanged
10 Diameter at hounds	5½in (13.97cm) Assumed	Unchanged	Unchanged
11 Topmast (square)	4⅛in (10.48cm) Assumed	Unchanged	Unchanged
12 Diameter at truck	2¾in (6.99cm) Assumed	Unchanged	Unchanged
FORE TOPGALLANT MAST (F)			
1 Length overall	35ft 7½in (10.854m)	25ft 10in (7.874m)	25ft 10in (7.874m)
2 Topgallant mast	23ft 5½in (7.150m)	17ft 3in (5.258m)	17ft 3in (5.258m)
3 Head (pole head)	12ft 2in (3.708m)	8ft 7in (2.616m)	8ft 7in (2.616m)
4 Top of cap to hounds	Unknown	12ft 2½in (3.721m)	Unchanged
5 Heeling to top of cap	Unknown	2ft 8in (81.28cm) Assumed	Unchanged
6 Heeling	Unknown	1ft 4in (40.64cm) Assumed	Unchanged
7 Hounds	Unknown	12½in (31.75cm) Assumed	Unchanged
8 Given diameter	6in (15.24cm)	4¾in (12.07cm) Assumed	Unchanged
9 Heeling (square)	Unknown	5¾in (14.68) Assumed	Unchanged
10 Diameter at hounds	Unknown	3⅞in (9.84cm) Assumed	Unchanged
11 Topmast (square)	Unknown	2⅞in (7.18cm) Assumed	Unchanged
12 Diameter at truck	Unknown	2in (5.08cm) Assumed	Unchanged
MAIN TOPGALLANT MAST (F)			
1 Length overall	34ft 7in (10.541m)	27ft 3in (8.304m)	27ft 3in (8.304m)

Continued overleaf

Item	As purchased with *Bethia*	As ordered by the Admiralty	As requested by Bligh
Main topgallant mast (F) continued			
2 Topgallant mast	23ft 10in (7.264m)	18ft 2in (5.537m)	18ft 2in (5.537m)
3 Head (pole head)	10ft 9in (3.277m)	9ft 1in (2.769m)	9ft 1in (2.769m}
4 Top of cap to hounds	Unknown	13ft 4in (4.064m) Assumed	Unchanged
5 Heeling to top of cap	Unknown	3ft 5in (1.041m) Assumed	Unchanged
6 Heeling	Unknown	1ft 4in (40.64cm) Assumed	Unchanged
7 Hounds	Unknown	1ft 5in (43.18) Assumed	Unchanged
8 Given diameter	6in (15.24cm)	6in (15.24cm)	Unchanged
9 Heeling (square)	Unknown	6¾in (17.15cm) Assumed	Unchanged
10 Diameter at hounds	Unknown	4⅞in (12.38cm) Assumed	Unchanged
11 Topmast (square)	Unknown	3in (7.62cm) Assumed	Unchanged
12 Diameter at truck	Unknown	2in (5.08cm)	Unchanged
FORE LOWER YARD (G)			
1 Length	40ft 5in (12.319m)	Unchanged	38ft 0in (11.582m)
2 Centre section	Unknown	Unknown	9ft 6in (2.896m) Assumed
3 Boom irons	Unknown	Unknown	6ft 4in (1.93m) Assumed
4 Gap at sling cleats	Unknown	Unknown	17¾in (45.08cm) Assumed
5 Sling cleat	Unknown	Unknown	11in (27.34cm) Assumed
6 Yard arm	Unknown	Unknown	1ft 7in (48.26cm)
7 Yard arm cleat	Unknown	Unknown	4½in (11.43cm) Assumed
8 Given diameter	11in (27.34cm)	10in (25.4cm)	8⅞in (22.54cm)
9 Diameter at yard arm	Unknown	Unknown	3⅞in (9.72cm) Assumed
MAIN LOWER YARD (G)			
1 Length	40ft 5in (12.206m)	46ft 6in (14.173m)	41ft 0in (12.497m)
2 Centre section	Unknown	Unknown	10ft 3in (3.124m) Assumed
3 Boom irons	Unknown	Unknown	6ft 10in (2.083m) Assumed
4 Gap at sling cleats	Unknown	Unknown	1ft 7in (48.26cm) Assumed
5 Sling cleats	Unknown	Unknown	11⅞in (30.16cm)
6 Yard arm	Unknown	Unknown	1ft 8½in (52.77cm) Assumed
7 Yard arm cleat	Unknown	Unknown	4¾in (12.07cm) Assumed
8 Given diameter	11¼in (28.58cm)	11¾in (29.85cm)	9½in (24.13cm)
9 Diameter at yard arm	Unknown	Unknown	4in (10.16cm)
MIZZEN YARD (GAFF) (J)			
1 Length	21ft 2in (6.452m)	21ft 2in (6.452m)	21ft 2in (6.452m)
2 Jaws	Unknown	Unknown	3ft 3in (99.76cm) Assumed
3 Given diameter	Unknown	Unknown	6⅛in (15.56cm) Assumed
4 Diameter at outer end	Unknown	Unknown	3⅝in (9.2cm) Assumed
CROSS JACK YARD (I)			
1 Length	36ft 2in (11.024m)	30ft 6in (9.926m)	30ft 6in (9.926m)
2 Centre section	Unknown	Unknown	7ft 7½in (2.324m) Assumed
3 Gap at sling cleats	Unknown	Unknown	6½in (16.51cm) Assumed
4 Sling cleats	Unknown	Unknown	8⅛in (20.64cm)
5 Yard arm	Unknown	Unknown	1ft 3¼in (38.74cm) Assumed
6 Yard arm cleat	Unknown	Unknown	2⅝in (6.67cm) Assumed
7 Given diameter	7½in (19.75cm)	6½in (16.51cm)	7½in (16.51cm)
8 Diameter at yard arm	Unknown	Unknown	2¾in (6.99cm) Assumed
FORE TOPYARD (H)			
1 Length	30ft 6in (9.3m)	30ft 6in (9.3m)	28ft 0in (8.534m)
2 Centre section	Unknown	Unknown	7ft 0in (2.134m)
3 Boom irons	Unknown	Unknown	4ft 7in (1.4m) Assumed
4 Gap at sling cleats	Unknown	Unknown	11¾in (29.84cm) Assumed
5 Sling cleat	Unknown	Unknown	7⅜in (18.73cm)
6 Yard arm	Unknown	Unknown	14in (35.56cm) Assumed

Item	As purchased with *Bethia*	As ordered by the Admiralty	As requested by Bligh
Fore topyard (H) continued			
7 Yard arm cleat	Unknown	Unknown	2⅞in (7.3cm) Assumed
8 Given diameter	7¾in (19.64cm)	7in (7.78cm)	5⅞in (14.43cm)
9 Diameter at yard arm	Unknown	Unknown	2½in (6.35cm) Assumed
MAIN TOPYARD (H)			
1 Length	35ft 2in (2.719m)	35ft 2in (2.719m)	31ft 0in (9.444m)
2 Centre section	Unknown	Unknown	7ft 9in (2.362m) Assumed
3 Boom irons	Unknown	Unknown	5ft 2in (1.575m) Assumed
4 Gap at sling cleats	Unknown	Unknown	13in (33.72cm) Assumed
5 Sling cleat	Unknown	Unknown	8⅛in (20.64cm) Assumed
6 Yard arm	Unknown	Unknown	15½in (39.37cm) Assumed
7 Yard arm cleat	Unknown	Unknown	3⅛in (5.25cm) Assumed
8 Given diameter	8 1/8in (20.64cm)	8⅛in (20.64cm)	6⅛in (15.51cm)
9 Diameter at yard arm	Unknown	Unknown	2¾in (6.98cm)
MIZZEN TOPYARD (I)			
1 Length	24ft 0in (7.315m)	24ft 0in (7.315m)	24ft 0in (7.315m)
2 Centre section	Unknown	Unknown	6ft 0in (1.829m)
3 Gap at sling cleats	Unknown	Unknown	10in (25.4cm)
4 Sling cleat	Unknown	Unknown	6¼in (15.87cm) Assumed
5 Yard arm	Unknown	Unknown	1ft 6in (45.72cm) Assumed
6 Yard arm cleat	Unknown	Unknown	2¼in (5.71cm) Assumed
7 Given diameter	5¼in (13.33cm)	5in (12.7cm)	5in (12.7cm)
8 Diameter at yard arm	Unknown	Unknown	2¼in (5.71cm) Assumed
FORE TOPGALLANT YARD (I)			
1 Length	Unknown	24ft 8⅝in (7.534m)	24ft 8in (7.417m)
2 Centre section	Unknown	Unknown	6ft 2in (1.879m) Assumed
3 Gap at sling cleats	Unknown	Unknown	10in (25.4cm) Assumed
4 Sling cleat	Unknown	Unknown	6¼in (15.87cm) Assumed
5 Yard arm	Unknown	Unknown	2ft 0½in (62.23cm) Assumed
6 Yard arm cleat	Unknown	Unknown	2½in (6.35cm)
7 Given diameter	Unknown	5in (12.7cm)	5in (12.7cm)
8 Diameter at yard arm	Unknown	Unknown	2⅛in (5.4cm) Assumed
MAIN TOPGALLANT YARD (I)			
1 Length	Unknown	24ft 8⅝in (7.534m)	24ft 8in (7.417m)
2 Centre section	Unknown	Unknown	6ft 2in (1.879m)
3 Gap at sling cleats	Unknown	Unknown	10in (25.4cm) Assumed
4 Sling cleat	Unknown	Unknown	6¼in (15.87cm) Assumed
5 Yard arm	Unknown	Unknown	2ft 0½in (62.23cm) Assumed
6 Yard arm cleat	Unknown	Unknown	2½in (6.35cm) Assumed
7 Given diameter	Unknown	5in (12.7cm)	5in (12.7cm)
8 Diameter at yard arm	Unknown	Unknown	2⅛in (5.4cm} Assumed
SPRITSAIL YARD (I)			
1 Length	Unknown	30ft 6in (9.296m)	30ft 6in (9.296m)
2 Centre section	Unknown	Unknown	7ft 7½in (2.324m)
3 Gap at sling cleats	Unknown	Unknown	6in (15.24cm) Assumed
4 Sling cleat	Unknown	Unknown	7.5in (19.05cm) Assumed
5 Yard arm	Unknown	Unknown	15½in (39.37cm) Assumed
6 Yard arm cleat	Unknown	Unknown	3in (7.62cm) Assumed
7 Given diameter	Unknown	6⅜in (16.19cm)	6in (15.24cm)
8 Diameter at yard arm	Unknown	Unknown	2⅝in (6.67cm)

Continued overleaf

Item	As purchased with *Bethia*	As ordered by the Admiralty	As requested by Bligh
FORE LOWER STUDDING SAIL BOOM (K)			
1 Length	Unknown	Unknown	21ft 1in (6.426m) Assumed
2 Taper	Unknown	Unknown	14ft 1in (4.292m) Assumed
3 Straight	Unknown	Unknown	7ft 0in (2.134) Assumed
4 Given diameter	Unknown	Unknown	4¼in (10.73cm) Assumed
5 Diameter at outer end	Unknown	Unknown	2⅞in (7.3cm) Assumed
MAIN LOWER STUDDING SAIL BOOM (K)			
1 Length	Unknown	Unknown	22ft 9in (6.934m) Assumed
2 Taper	Unknown	Unknown	15ft 2in (4.623m) Assumed
3 Straight	Unknown	Unknown	7ft 7in (2.311m) Assumed
4 Given diameter	Unknown	Unknown	4½in (11.43cm) Assumed
5 Diameter at outer end Assumed	Unknown	Unknown	3in (7.62cm)
FORE TOP STUDDING SAIL BOOM (L)			
1 Length	Unknown	Unknown	19ft 0in (5.791m) Assumed
2 Straight	Unknown	Unknown	6ft 4in (1.931m) Assumed
3 Taper	Unknown	Unknown	12ft 8in (3.861m)
4 Given diameter	Unknown	Unknown	3⅞in (9.842cm) Assumed
5 Diameter at outer end	Unknown	Unknown	2⅝in (6.67cm) Assumed
MAIN TOP STUDDING SAIL BOOM (L)			
1 Length	Unknown	Unknown	20ft 6in (6.248m) Assumed
2 Straight	Unknown	Unknown	6ft 10in (2.083m) Assumed
3 Taper	Unknown	Unknown	13ft 8in (4.166m) Assumed
4 Given diameter	Unknown	Unknown	4⅛in (10.48cm) Assumed
5 Diameter at outer end	Unknown	Unknown	2¾in (6.98cm) Assumed
FORE TOPGALLANT STUDDING SAIL BOOM (L)			
1 Length	Unknown	Unknown	14ft 0in (4.267m) Assumed
2 Straight	Unknown	Unknown	4ft 8in (1.422m) Assumed
3 Taper	Unknown	Unknown	9ft 4in (2.844m) Assumed
4 Given diameter	Unknown	Unknown	2⅞in (7.17cm) Assumed
5 Diameter at outer end	Unknown	Unknown	2in (5.78cm) Assumed
MAIN TOPGALLANT STUDDINGG SAIL BOOM (L)			
1 Length	Unknown	Unknown	15ft 6in (4.724m) Assumed
2 Straight	Unknown	Unknown	5ft 2in (1.575m) Assumed
3 Taper	Unknown	Unknown	10ft 4in (3.148m) Assumed
4 Given diameter	Unknown	Unknown	1⅜in (8.67cn) Assumed
5 Diameter at outer end	Unknown	Unknown	2in (5.78cm) Assumed
FORE STUDDING SAIL YARD (M)			
1 Length	Unknown	Unknown	12ft 0in (3.658m) Assumed
2 Outer arm (Tapered)	Unknown	Unknown	8ft 0in (2.433m) Assumed
3 Inner arm (Tapered)	Unknown	Unknown	4ft 0in (1.213m) Assumed
4 & 5 Sling cleats & gap	Unknown	Unknown	2½in (6.35cm) Assumed
6 Yard arm	Unknown	Unknown	5in (12.7cm) Assumed
7 Given diameter	Unknown	Unknown	2½in (6.35cm) Assumed

Item	As purchased with *Bethia*	As ordered by the Admiralty	As requested by Bligh
Fore studding sail yard (M) continued			
8 Diameter at outer ends	Unknown	Unknown	1$\frac{5}{8}$in (4.13cm) Assumed
MAIN STUDDING SAIL YARD (M)			
1 Length	Unknown	Unknown	13ft 0in (3.962cm) Assumed
2 Outer arm (tapered)	Unknown	Unknown	8ft 8in (2.64m) Assumed
3 Inner arm (tapered)	Unknown	Unknown	4ft 4in (1.321m) Assumed
4 & 5 Sling cleats & gap	Unknown	Unknown	3in (7.62cm) Assumed
6 Yard arm	Unknown	Unknown	9in (22.86cm) Assumed
7 Given diameter	Unknown	Unknown	4$\frac{1}{2}$in (11.43cm) Assumed
8 Diameter at outer ends	Unknown	Unknown	1 3.4in (4.44cm) Assumed
FORE TOP STUDDING SAIL YARD (M)			
1 Length	Unknown	Unknown	10ft 10in (3.302m) Assumed
2 Outer arm (tapered)	Unknown	Unknown	7ft 3in (2.21m) Assumed
3 Inner arm (tapered)	Unknown	Unknown	3ft 7in (1.092m) Assumed
4 & 5 Sling cleats & gap	Unknown	Unknown	2in (5.08cm) Assumed
6 Yard arm	Unknown	Unknown	4$\frac{1}{8}$in (10.48cm) Assumed
7 Given diameter	Unknown	Unknown	2$\frac{1}{8}$in (5.4cm) Assumed
8 Diameter at outer ends	Unknown	Unknown	1$\frac{1}{2}$in (3.81cm) Assumed
MAIN TOP STUDDING SAIL YARD (M)			
1 Length	Unknown	Unknown	11ft 9in (3.581m) Assumed
2 Outer arm (tapered)	Unknown	Unknown	7ft 10in (2.387m) Assumed
3 Inner arm (tapered)	Unknown	Unknown	3ft 11in (1.194m) Assumed
4 &5 Sling cleats & gap	Unknown	Unknown	2$\frac{1}{2}$in 6.35cm) Assumed
6 Yard arm	Unknown	Unknown	4$\frac{3}{4}$in (12.06cm) Assumed
7 Given diameter	Unknown	Unknown	2$\frac{3}{8}$in (6.03m) Assumed
8 Diameter at outer ends	Unknown	Unknown	1$\frac{1}{2}$in (3.81cm) Assumed
FORE TOPGALLANT STUDDING SAIL YARD (M)			
1 Length	Unknown	Unknown	8ft 0in (2.433m) Assumed
2 Outer arm (tapered)	Unknown	Unknown	5ft 4in (1.626m) Assumed
3 Inner arm (tapered)	Unknown	Unknown	2ft 8in (81.28cm) Assumed
4 & 5 Sling cleats & gap	Unknown	Unknown	1$\frac{5}{8}$in (4.13cm) Assumed
6 Yard arm	Unknown	Unknown	3$\frac{1}{4}$in (8.25cm) Assumed
7 Given diameter	Unknown	Unknown	1$\frac{5}{8}$in (4.13cm) Assumed
8 Diameter at outer ends	Unknown	Unknown	1$\frac{1}{8}$in (2.86cm) Assumed
MAIN TOPGALLANT STUDDING SAIL YARD (M)			
1 Length	Unknown	Unknown	8ft 10$\frac{1}{4}$in (2.699m) Assumed
2 Outer arm (tapered)	Unknown	Unknown	5ft 10$\frac{3}{4}$in (5.896m) Assumed
3 Inner arm (tapered)	Unknown	Unknown	2ft 11$\frac{1}{2}$in (92.71cm) Assumed
4 & 5 Sling cleats & gap	Unknown	Unknown	2in (5.78cm) Assumed
6 Yard arm	Unknown	Unknown	3$\frac{3}{4}$in (9.52cm) Assumed
7 Given diameter	Unknown	Unknown	1$\frac{7}{8}$in (4.76cm) Assumed
8 Diameter at outer ends	Unknown	Unknown	1$\frac{1}{4}$in (3.17cm) Assumed

Note: Dimensions compiled by John McKay.

Selected References

BIBLIOGRAPHY

Alexander, C, *The Bounty* (Penguin, New York 2003).

Bligh, W, *A voyage to the South Sea undertaken by Command of His Majesty for the purpose of conveying the bread-fruit tree to the West Indies in His Majesty's Ship the Bounty commanded by Lieutenant William Bligh and an account of the mutiny on board H.M.S. Bounty and the subsequent voyage of part of the crew in the ship's boat, from Tofoa, one of the Friendly Islands, to Timor, a Dutch settlement in the East Indies* (originally published in 1792; paperback edition Signet, New York 1962).

Bligh, W, *Log of the Proceedings of His Majestys Ship Bounty Lieut. Wm Bligh Commander from Otaheite towards Jamaica, signed 'Wm Bligh', (5 April 1789 – 13 March 1790)* (State Library of New South Wales, Sydney; facsimile).

Dening, G, *Mr Bligh's Bad Language: Passion, Power and Theatre on the Bounty* (Cambridge University Press, Cambridge 1992).

Franklin, John, *Navy Board Ship Models 1650–1750* (Conway Maritime Press, London & Naval Institute Press, Annapolis 1989).

Harland, John H, *Seamanship in the Age of Sail: An Account of the Shiphandling of the Sailing Man of-war, 1600-1860, Based on Contemporary Sources* (Conway Maritime Press, London 1984).

Goodwin, Peter, 'The Application and Scheme of Paintworks in British Men-of-War in the Late Eighteenth and Early Nineteenth Centuries', *The Mariner's Mirror*, 99(3), (2013), pp287-300. DOI: 10.1080/00253359.2013.815993.

Goodwin, Peter, *The Construction and Fitting of the English Man of War 1650–1850* (Conway Maritime Press, London & Naval Institute Press, Annapolis 1987).

Hough, Richard, *Captain Bligh and Mr Christian* (Hutchinson, London 1972).

Jang, Kerry, *Ship Models from the Age of Sail: Building and Enhancing Commercial Kits* (Seaforth Publishing, Barnsley 2022).

Knight, C, 'The Armed Vessel Bounty', *The Mariner's Mirror*, 22(2), (1936), pp183-199. DOI: 10.1080/00253359.1936.10657185

McKay, John, *The Armed Transport Bounty* (Conway Maritime Press, London 1989).

Mansir, A R, *The Journal of Bounty's Launch* (Kittiwake, Montrose, 1989).

Morrissey, M, *Breadfruit Stories. Trees That Feed* (Winneteka 2021).

Petersson, L, *Rigging Period Ship Models* (Chatham Publishing, London 2000).

Platt, R, & Beisty, S, *Stephen Biesty's Cross-Sections: Man-of-War* (Dorling Kindersley, London 1993).

Pulvertaft, D, 'The Colour Schemes of British Warship Figureheads 1727–1900', *The Mariner's Mirror*, 104:2, (2018), pp192-210, DOI: 10.1080/00253359.2018.1453650

Roberts, D H (ed), *18th Century Shipbuilding: Remarks on the navies of the English and Dutch by Blaise Ollivier (1737)* (Jean Boudriot, Paris 1992).

Thomas, N, 'Domestic structures and polyandry in the Marquesas Islands', pp65-83, in Jolly, M & MacIntyre, M (eds), *Family and Gender in the Pacific: Domestic Contradictions and the Colonial Impact* (Cambridge University Press, Cambridge 1989).

Vale, B, 'Pitch, Paint, Varnish and the Changing Colour Schemes of Royal Navy Warships,1775–1815: A summary of existing knowledge', *The Mariner's Mirror*, 106(1), (2020), pp30-42. DOI: 10.1080/00253359.2020.1712841

Reed, P, *Modelling Sailing Men-of-War* (Chatham Publishing, Rochester & Naval Institute Press, Annapolis 2000).

Reed, P, *Period Ship Model Making* (Seaforth Publishing, Barnsley & Naval Institute Press, Annapolis 2007).

Walters, S S (ed), *The Voyage of the Bounty Launch: John Fryer's Narrative* (Genesis Publications. Guildford 1979).

ONLINE RESOURCES

Research

Copies of John McKay's *Bounty* drawings can be obtained by writing to him at: johnwmckay@telus.net

Royal Museum Greenwich Archives
https://www.rmg.co.uk/collections/

Nautical Research Guild
https://www.thenrg.org/

Model Ship World
https://modelshipworld.com/

Links to *Bounty* Kits and Accessories

HiSModel
https://www.hismodel.com/

DeAgostini
https://www.deagostini.com/

Occre
https://occre.com/

Amati
https://store.amatimodel.com/

Dusek/Mamoli
http://www.dusekshipkits.com/

Model Expo
https://modelexpo-online.com/

The Lumberyard
https://www.dlumberyard.com/

Modelship Builder (Hahn Plans)
http://www.modelshipbuilder.com/

Cornwall Model Boats
https://www.cornwallmodelboats.co.uk/

Hannants
https://www.hannants.co.uk/